IAN RAMSEY CENTRE STUDIES IN SCIENCE AND RELIGION

General Editor: ALISTER E. MCGRATH
Managing Editor: ANDREW PINSENT

The *Ian Ramsey Centre Studies in Science and Religion* series brings readers innovative books showcasing cutting-edge research in the field of science and religion. The series will consider key questions in the field, including the interaction of the natural sciences and the philosophy of religion; the impact of evolutionary theory on our understanding of human morality, religiosity, and rationality; the exploration of a scientifically-engaged theology; and the psychological examination of the importance of religion for human flourishing and well-being. The series will also encourage the development of new and more nuanced readings of the interaction of science and religion. This ground-breaking series aims to represent the best new scholarship in this ever-expanding field of study.

The Territories of Human Reason

*Science and Theology in an Age
of Multiple Rationalities*

ALISTER E. McGRATH

OXFORD
UNIVERSITY PRESS

Great Clarendon Street, Oxford, OX2 6DP,
United Kingdom

Oxford University Press is a department of the University of Oxford.
It furthers the University's objective of excellence in research, scholarship,
and education by publishing worldwide. Oxford is a registered trade mark of
Oxford University Press in the UK and in certain other countries

© Alister E. McGrath 2019

The moral rights of the author have been asserted

First Edition published in 2019

Impression: 1

All rights reserved. No part of this publication may be reproduced, stored in
a retrieval system, or transmitted, in any form or by any means, without the
prior permission in writing of Oxford University Press, or as expressly permitted
by law, by licence or under terms agreed with the appropriate reprographics
rights organization. Enquiries concerning reproduction outside the scope of the
above should be sent to the Rights Department, Oxford University Press, at the
address above

You must not circulate this work in any other form
and you must impose this same condition on any acquirer

Published in the United States of America by Oxford University Press
198 Madison Avenue, New York, NY 10016, United States of America

British Library Cataloguing in Publication Data

Data available

Library of Congress Control Number: 2018952215

ISBN 978-0-19-881310-1

Printed and bound in Great Britain by
Clays Ltd, Elcograf S.p.A.

In memory of
Ian T. Ramsey (1915–72)
Nolloth Professor of the Philosophy of the Christian
Religion, Oxford University
Bishop of Durham

Contents

Introduction: Science and Theology in an Age of 'Multiple
   Situated Rationalities' — 1
Mapping the Territories of Human Reason — 3
Mapping the Territories of Science and Religion — 6
The Aim of this Book — 13

I. EXPLORING THE NOTION OF RATIONALITY

1. One Reason; Multiple Rationalities: The New Context
   of Discussion — 19
   Shifting Notions of Rationality — 21
   Rationality, Embodiment, and Embeddedness — 22
   Reflections on the Cultural and Social Embeddedness of Rationality — 27
   The Embodiment of Right Reason: The 'Wise' — 32
   Concerns about Human Rationality — 35
   One Reason; Multiple Rationalities — 39
   Rationality, Ideology, and Power — 46

2. Mapping Human Reason: Rationalities across
   Disciplinary Boundaries — 50
   On the Correlation of Rationalities — 53
   Scientism: The Natural Sciences as the Ultimate Rational Authority — 56
   Multiple Perspectives on a Complex Reality — 59
   Science and Theology: Distinct Perspectives on Reality — 63
   Science and Theology: Distinct Levels of Reality — 65

3. Social Aspects of Rationality: Tradition and Epistemic
   Communities — 75
   Communities and their Epistemic Systems — 76
   Rationality, Community, and Tradition — 80
   Rationality and Dominant Cultural Metanarratives — 84
   Science and Religion: Reflections on the Communal
      Aspects of Knowledge — 89

II. RATIONALITY IN SCIENCE AND THEOLOGY: A CRITICAL ENGAGEMENT

4. Rational Virtues and the Problem of Theory Choice — 95
   What is a Theory? — 97
   Inference to the Best Explanation — 101
   Correspondence and Coherence as Theoretical Virtues — 106
   Objectivity — 110
   Simplicity — 113
   Elegance and Beauty — 117
   A Capacity to Predict — 119

5. Rational Explanation in Science and Religion — 124
   What does it Mean to 'Explain'? — 125
   Causality as Explanation — 129
   Unification as Explanation — 132
   Two Approaches to Explanation: Ontic and Epistemic — 135
   Religious Explanation: Some General Reflections — 137
   Religious Explanation: Ontic and Epistemic — 140
   Theology, Ontology, and Explanation — 143
   A Case Study: Aquinas's 'Second Way' — 145
   The Image of God and Religious Explanation — 148
   Understanding and Explaining: A Religious Perspective — 150

6. From Observation to Theory: Deduction, Induction, and Abduction — 154
   The Entanglement of Theory and Observation — 157
   Logics of Discovery and Justification — 159
   Deduction in the Natural Sciences — 164
   Deduction in Christian Theology — 167
   Induction in the Natural Sciences — 170
   Induction in Christian Theology — 172
   Abduction in the Natural Sciences — 175
   Abduction in Christian Theology — 178

7. Complexity and Mystery: The Limits of Rationality in Science and Religion — 182
   Mystery and Irrationality — 182
   Mystery in Science — 187
   Mystery in Christian Theology — 190
   The Trinity as Mystery — 195

Contents ix

On Being Receptive to Mystery	199
Mystery: An Invitation to Deeper Reflection	201
8. Rational Consilience: Some Closing Reflections on Science and Christian Theology	203
Towards a 'Big Picture': A Metaphysical Turn	204
The Colligation of Insights	210
A Case Study in Colligation: Science and Socialism	212
A Case Study in Colligation: Science and Theology	218
Rationality: A Cohesive Approach	221
Bibliography	227
Index	283

Introduction

Science and Theology in an Age of 'Multiple Situated Rationalities'

Philip Clayton opens his 1989 study of the concept of explanation in physics and theology with an arresting and engaging statement: 'For believers, religious beliefs help to explain the world and their place within it.'[1] For Clayton, this represents both a reliable summary of the consensus of religious believers, and a legitimate option within the changing intellectual landscape of that age. Clayton rightly emphasizes the radical changes in scholarly understanding of rationality which lay behind his book. The school of logical positivism, dominant in the early 1930s and still influential in the 1960s, left no conceptual space for 'rational' discussion of beliefs about God, generally taking the view that religious language could not be cognitively meaningful. Yet major transformation in the philosophy of science began to take place during the 1950s, as positivist accounts of reality were gradually displaced by contextualist or coherence-based theories of scientific rationality, opening up new possibilities of dialogue between theology and the philosophy of science.[2]

This book seeks to extend this discussion in the light of still further changes in the wider debate about human rationality, including but extending beyond the philosophy of science, since the publication of

---

[1] Clayton, *Explanation from Physics to Theology*, 1. While Clayton notes additional functions of religious belief, he rightly insists that a capacity to explain remains an integral aspect of faith. Cf. van Huyssteen, *Essays in Postfoundationalist Theology*, 231.

[2] For some important episodes in the history of exploration of this theme, see Zachhuber, *Theology as Science in Nineteenth-Century Germany*.

Clayton's landmark book—above all, the seemingly inexorable move away from the notion of a single universal rationality towards a plurality of cultural and domain-specific methodologies and rationalities. In particular, it seeks to affirm and explore the intellectual legitimacy of both interdisciplinary and transdisciplinary dialogue[3] in this age of multiple situated rationalities, focussing on the interaction of the natural sciences and Christian theology as a single case study with the potential to illuminate other such discussions.

The suggestion that the natural sciences themselves adopt a plurality of methods and criteria of rationality finds ample support in scientific practice. The biologist Steven Rose, reflecting on the complexity of the scientific task of engaging and explaining the world, drew a conclusion he believed to be widely shared among reflective scientists. 'As a materialist, as all biologists must be, I am committed to the view that we live in a world that is an *ontological unity*, but I must also accept an *epistemological pluralism*.'[4] We cannot reduce all cognitive activity to 'a single fundamental method', as some scientists suggest,[5] but must rather make use of a range of conceptual tool-boxes, adapted to specific tasks and situations, to give as complete an account as possible of our world.[6] Different explanatory types or 'stances' are evident across disciplines—an observation which has led to a sea change within the philosophy of science, which has moved away from older reductionist or eliminativist views of explanation in

---

[3] The distinction between 'interdisciplinary' and 'transdisciplinary' is contested. A common way of understanding the relation between multidisciplinary, interdisciplinary, and transdisciplinary approaches is to see them respectively as additive, interactive, and holistic. On this way of thinking, interdisciplinarity aims to create interactive links among disciplines which allows them to be seen as a coordinated and coherent whole. Transdisciplinarity aims to integrate disciplinary insights and transcend their traditional boundaries, thus adding a new element to the discussion that cannot be sustained by any single discipline. For a discussion, see Leavy, *Essentials of Transdisciplinary Research*; Osborne, 'Problematizing Disciplinarity, Transdisciplinary Problematics'.

[4] Rose, 'The Biology of the Future and the Future of Biology', 128–9 (my emphasis).

[5] See, for example, Ladyman and Ross, *Every Thing Must Go*; Rosenberg, *The Atheist's Guide to Reality*. For a critique of such a scientific exclusivism, see Kidd, 'Reawakening to Wonder'.

[6] Gigerenzer, 'The Adaptive Toolbox', 38–43. For a more philosophical approach, see Midgley, *The Myths We Live By*, 76–7.

Introduction 3

favour of pluralistic approaches.[7] For many, this line of thought leads ultimately to the conclusion that the 'only route to a deeper understanding of ourselves is through radical epistemological pluralism'.[8] Yet while the pluralistic nature of scientific enquiry seems clear to anyone acquainted with its history or practice, the implications of this pluralism require further exploration.

MAPPING THE TERRITORIES OF HUMAN REASON

This book offers a tentative and provisional mapping of human rationality, surveying the forms of reasoning and criteria of rationality that have characterized the production of knowledge across culture and history, and within specific disciplines. The failings of certain approaches which emerged during the Enlightenment—specifically, a family of phenomenological views that saw human reason as historically and culturally invariant, or ideological views that took the approaches to reason developed by Western European *illuminati* of the eighteenth century as normative—are now seen to depend on unreliable mappings of the multiple territories of human reason. There is no universal 'republic of reason'—rather, we have to contend with an array of distinct, yet occasionally overlapping and competing, epistemic territories and communities. This has led to growing interest in transdisciplinarity—the quest for 'articulated conceptual frameworks' that transcend the narrow scope of disciplinary worldviews, and thus offer an enriched and deepened understanding of our world.[9]

---

[7] Keil, 'Explanation and Understanding', 231–3.

[8] Dupré, 'The Lure of the Simplistic', 293. There is no need to follow Paul Feyerabend and move from the recognition of a plurality of methods in science to a form of relativism that would be seen by many, if not most, scientists as ultimately compromising the integrity of science: Pigliucci, 'Feyerabend and the Cranks'; Kidd, 'Why Did Feyerabend Defend Astrology?'

[9] Klein, 'Discourses of Transdisciplinarity'. The emergence of transdisciplinarity is usually dated to a conference held in 1970 at the University of Nice by the Organization of Economic Cooperation and Development (OECD), which recognized the need for 'a common system of axioms for a set of disciplines' that transcends the narrow scope of disciplinary worldviews through an overarching synthesis. The term

In exploring the multiple territories of human reason, this work aims to contribute to discussions about the possibilities of interdisciplinary and transdisciplinary discourse and reflection, with the object of enriching knowledge and understanding by avoiding disciplinary compartmentalization and encouraging a transdisciplinary inquiry characterized by a common orientation to transcend disciplinary boundaries and an attempt to bring continuity to enquiry and knowledge through attention to comprehensiveness and creating porous boundaries between concepts and disciplines. While this question is of general interest, it is of especial interest to the rapidly growing field of 'science and religion', which has established itself since its emergence in the 1960s as one of the most interesting, challenging, and contested areas of academic investigation and reflection.

The philosopher John Dewey famously argued that the 'deepest problem of modern life' was our collective and individual failure to integrate our 'thoughts about the world' with our thoughts about 'value and purpose'.[10] Dewey's distinct proposal for the redirection of philosophy, bringing it back into contact with the deep human concerns from which he believed it had originally sprung, may have been relegated to the domain of intellectual history; the issue he raised, however, has never gone away. Others have echoed his concern about divorcing philosophy from life's deepest questions and concerns, suggesting that a purely 'objective' account of the world disregards most of what makes life interesting.[11] Science 'may provide the most useful way to organize empirical, reproducible data', but its power to do so is 'predicated on its inability to grasp the most central aspects of human life'—such as love, beauty, honour, and virtue.[12]

Yet how can we follow Dewey in aiming to integrate such thoughts about the world, value, and practice, while acting *rationally* throughout this process? The empirical investigation of our world is the domain of the natural sciences, which some hold to be characterized

---

'transdisciplinary' is widely attributed to Jean Piaget: see, for example, López-Huertas, 'Reflections on Multidimensional Knowledge'.

[10] Dewey, *The Quest for Certainty*, 255. On Dewey, see Kitcher, 'The Importance of Dewey for Philosophy (and for Much Else Besides)'.

[11] Davies, *Why Beliefs Matter*, v.

[12] Kalanithi, *When Breath Becomes Air*, 151.

by a rational precision which is quite absent from any exploration of non-empirical notions, such as value, meaning, or purpose—notions that are traditionally seen as falling within the domain of religion, which provides a way of helping individuals to transcend their own concerns or experience and connect up with something greater.[13] Mary Midgley, for example, suggests that science 'only begins to have a value' when it is 'brought into contact with some existing system of aims and purposes'.[14] To give a full account of our complex world, enabling us to live meaningfully within it, requires that we develop a correspondingly complex array of research methods, disciplines, and traditions for making sense of it. It also requires a discourse of transgression—a principled belief that disciplinary boundaries can be a barrier to a deeper form of knowledge, and the development of transdisciplinary strategies to allow discourse across and within those boundaries.[15]

In famously declaring that 'a *picture* held us captive',[16] making it difficult for us to liberate ourselves from its imaginative thrall, Ludwig Wittgenstein was basically pointing out how easily our understanding of our world can be controlled by an 'organizing myth'[17]—a worldview or metanarrative that has, whether we realize it or not, come to dominate our perception of our world, in effect predisposing us to interpret experience in certain manners as natural or self-evidentially correct, while blinding us to alternative ways of understanding it. We are easily seduced by simplistic, captivating narratives about our world, coming to believe we are seeing things 'the way they really are', when we are in fact seeing them in a specific way, which is often

---

[13] Park, 'Religion as a Meaning-Making Framework in Coping with Life Stress'; Emmons, *The Psychology of Ultimate Concerns*; Wong, *The Human Quest for Meaning*. There is, of course, continuing interest in such concepts within philosophy: see, for example, Wolf, *Meaning in Life*; Seachris, 'The Meaning of Life as Narrative'.

[14] Midgley, *The Myths We Live By*, 21. A similar theme is found throughout Wilson, *Consilience*.

[15] See especially Biagioli, 'Postdisciplinary Liaisons'; Osborne, 'Problematizing Disciplinarity, Transdisciplinary Problematics'. For a classic account of the deliberate attempt to justify and enforce disciplinary boundaries, see Gieryn, 'Boundary-Work and the Demarcation of Science from Non-Science'.

[16] Wittgenstein, *Philosophical Investigations*, §115.

[17] For this interpretation of Wittgenstein's use of *Bild*, see Egan, 'Pictures in Wittgenstein's Later Philosophy'.

socially constructed rather than 'natural', and are thus failing to appreciate both the specificity and limits of this perspective, or the plausibility of alternative ways of framing our world. Wittgenstein's use of the analogy of a picture (*Bild*) aims to free us from the grip which *certain* pictures have on us,[18] especially by inviting us to view alternative paradigms or framings, which might offer a better rendering of our world.

Wittgenstein's point has particular force in the field of science and religion, in that the culturally regnant 'picture' of their relationship holds that they are in perennial and essential conflict.[19] Historical research has exposed the flaws of idealized, synthetic accounts of '*the* relationship between science and theology', making it clear that a multiplicity of such accounts may be given, and that the task of judging which is 'right' lies beyond normative definitive judgement on the basis of the resources available to us. Religion and science are both theoretical constructs, rather than natural types; both are shaped by cultural perceptions and agendas, which severely impede any attempt to offer essentialist accounts of their nature, or their possible interactions.[20] Yet the 'conflict myth' retains its appeal. Before proceeding further, we need to reflect on this point.

MAPPING THE TERRITORIES OF SCIENCE AND RELIGION

In a series of important and influential historical studies of science and religion in the 1990s and beyond, focussing especially on the nineteenth century, the Oxford scholar John Hedley Brooke has

---

[18] See the important essay in Baker, *Wittgenstein's Method*, 260–78.

[19] For the origins of the 'conflict' myth which underlies this influential yet unreliable mapping, see Russell, 'The Conflict Metaphor and Its Social Origins'; Turner, 'The Victorian Conflict between Science and Religion'; Watts, 'Are Science and Religion in Conflict?'; and especially Harrison, *The Territories of Science and Religion*, 172–6, 191–8.

[20] Hence the widespread scholarly demands to acknowledge that 'religion' (and 'non-religion') are categories of social designation which lack empirical or scientific warrant: see especially Jong, 'On (Not) Defining (Non)Religion'.

argued that serious scholarship in the history of science has revealed 'so extraordinarily rich and complex a relationship between science and religion in the past that general theses are difficult to sustain. The real lesson turns out to be the complexity'.[21] Brooke's analysis has found widespread support within the scholarly community. Peter Harrison has pointed out that 'study of the historical relations between science and religion does not reveal any simple pattern at all',[22] such as the monomyth of the 'conflict' narrative. It does, however, disclose a 'general trend'—that for most of the time, religion has *facilitated* scientific enquiry.

This scholarly complexification of the relation of science and religion has left many feeling 'emotionally and intellectually unsatisfied', in that it 'seems to have little to recommend it besides its truth'.[23] For some, any suggestion that science and religion might exist, even to a limited extent, in a non-confrontational relationship seems to cause cultural anxiety and unease in some,[24] perhaps reflecting a fear of intellectual contamination or category violation. Yet while this account may lack popular appeal or a simple rhetorical flourish, it has the enormous advantage of corresponding more closely to the historical evidence, and thus offers a more reliable basis for informed discussion of the issues.

Picking up on Wittgenstein's image of a limiting and imprisoning *Bild*, Peter Harrison has argued that the 'picture' of science and religion that has held Western culture captive for the last century is a flawed conceptual map that forces the disciplines of science and religion to be seen as existing in an agonistic relationship.[25] The perception thus becomes a reality. Yet whether science and religion are seen to be at war with each other depends on how their relationship is framed—on what map is used to identify their respective territories.

---

[21] Brooke, *Science and Religion*, 6. For Brooke's personal reflections on his historical research, see Brooke, 'Living with Theology and Science'.

[22] Harrison, 'Introduction', 4.

[23] Numbers, 'Simplifying Complexity', 263.

[24] For reflections on the grounds of such anxiety, see Russell, *Inventing the Flat Earth*, 35-49. There are interesting parallels here with the vigorous attempts in classical Greek culture to disentangle and disconnect science, mythology, magic, and philosophy: see Lloyd, *Magic, Reason and Experience*, 2-58.

[25] Harrison, *The Territories of Science and Religion*, 145-82.

A map represents a visual matrix developed to construct knowledge and meaning, combining various empirical perceptions and theoretical interpretations in 'mapping' a geographical, political, or intellectual territory[26]—and thus framing our perceptions of where something (such as 'science' or 'religion') properly belongs, and shaping our expectations of its relationships with other mapped objects. Whether science and religion are to be seen as in conflict depends on the categories and boundaries of their territories, which are social constructions, not empirical observations.[27]

Harrison's 'historical cartography of the categories of "religion" and "science"' both challenges the outdated 'warfare' map, and offers an alternative mapping of their respective territories which opens the way to a more reliable reading of intellectual history on the one hand, and present-day possibilities of interaction between science and religion on the other. Harrison's revisionist intellectual cartography helps to liberate us from becoming locked into an outdated narrative of conflict, by observing that what some deem to be intellectually necessary is culturally contingent—and that other more reliable cultural contingencies can be re-appropriated and renovated. Harrison's considerable intellectual achievement is to allow us to reorganize our conceptual domains to allow a fuller and more reliable appreciation of the situation.[28]

So given the ideological freighting of the 'conflict myth' and its highly unreliable historical foundations, why do this myth and its derived map persist in contemporary discussion? Harrison points out that this, like any other myth, serves an important function for certain communities of discourse—'validating a particular view of reality and a set of social practices'.[29] Given that the validity of certain particular views of reality—such as the specific forms of atheism associated with Richard Dawkins and Christopher Hitchens—has now become *dependent* upon this foundational narrative, there is no realistic possibility that it will be abandoned by such communities

---

[26] See Lévy, ed., *A Cartographic Turn*.
[27] See especially Harrison, '"Science" and "Religion": Constructing the Boundaries'.
[28] For the importance of such intellectual or imaginative reconfiguration, see Elgin, 'Creation as Reconfiguration'.
[29] Harrison, *The Territories of Science and Religion*, 173.

of discourse, whose existence depends upon its continuing plausibility. Paradoxically, this has become a sacred narrative for certain kinds of atheism, being treated as if it was immune to historical criticism and revision. My own experience of (non)-dialogue with such movements suggests that a compliant acquiescence in this manifestly flawed and discredited myth has become a boundary marker for inclusion within such communities.

Yet it is important to think, not merely in terms of the remapping of the domain of 'science and religion', but also of reconsidering the cultural mapping of the individual domains of both science and religion in their own right. In both cases, there are questions that need to be asked about popular perceptions and academic habits that clearly need careful and informed reconsideration. Some, for example, speak of the natural sciences' unique and characteristic use of the 'Scientific Method'. The fiction of a unique, singular, formalized set of methodological rules that constitutes 'the Scientific Method' (note the singular) lies behind much popular literature advocating a scientistic outlook, such as the writings of the Oxford physical chemist and popularizer Peter Atkins, who argues that the distinctive 'Scientific Method' is capable of illuminating 'every and any concept' in a uniquely reliable manner.[30] Yet this entertaining and simplistic account fails to take account of the distinct characteristics and objectives of individual sciences,[31] in effect reducing them all to a single 'mono-science' which overlooks their distinct identities, histories, and objects of enquiry.

This view of a unique 'Scientific Method' has long been undermined by scholarly studies of the history and practice of science, which point to a wide range of methods being deployed within the natural sciences, incapable of being reduced to a single 'method'. Perhaps the philosopher of science Paul Feyerabend overstated his case in *Against Method* (1975);[32] yet his critique of what we might style 'methodological monism' within the natural sciences remains on the table, along with its correlate—that there exists a plurality of

---

[30] Atkins, *On Being*, vii–ix.
[31] Clarke and Walsh, 'Scientific Imperialism and the Proper Relations between the Sciences'.
[32] For some important criticisms, see Preston, *Feyerabend*, 136–9; 174–7.

scientific *methods*, each adapted and developed in the light of a specific scientific discipline's investigative aims and objects of enquiry. Werner Heisenberg, in reflecting on the Copenhagen approach to quantum theory, highlighted the importance of one of its implications—that 'what we observe is not nature itself, but nature as it is disclosed by our methods of investigation'.[33] Even if we conceive nature as a unitary entity, Heisenberg's line of thought leads us to the conclusion that a multiplicity of research methods leads to a corresponding plurality of perspectives or insights, which thus require to be integrated, coordinated, or colligated in order to allow the best possible overall representation of nature.

The ontological unity of nature thus does not entail a single research method; rather, its depth and complexity demand an epistemological pluralism if it is to be fully and reliably characterized. No reliable map of intellectual territories can therefore be drawn on the basis of such a methodologically monistic notion of the natural or human sciences. This insight is fundamental to the approach adopted in this work, and will be developed further throughout its course.

But what of the domain of religion? It may be natural for us to think of religion in essentialist terms, seeing it as a universal category, embracing individual examples of this universal—such as Buddhism, Christianity, and Hinduism—so that summative generalizations can be made about the 'essence of religion'. Yet views about the nature, function, and identity of religion have varied from one historical location to another, as they do today. The category of 'religion' is best seen as a social construction that has little, if any, scientific legitimacy.[34] The term is socially important—for example, in relation to ensuring the basic right of 'religious freedom' (which requires some agreement on what counts as a religion.)[35] Yet such generally useful cultural conventions do not amount to a *scientific* understanding of religion, and must not be treated as such.

---

[33] Heisenberg, 'Die Kopenhagener Deutung der Quantentheorie', 85: 'Und wir müssen uns daran erinnern, daß das, was wir beobachten, nicht die Natur selbst ist, sondern Natur, die unserer Art der Fragestellung ausgesetzt ist.'

[34] See especially Jong, 'On (Not) Defining (Non)Religion', 16: 'our efforts to figure out how the term "religion" ought to be used have been—rightly or wrongly—motivated at least in part by a latent essentialism'.

[35] Zucca, 'A New Legal Definition of Religion?'

Introduction 11

As Jong points out, from an empirical standpoint, scholarly efforts to define 'religion' are confronted by two core difficulties—the 'Buddhism Problem' and the 'Football Problem'—which he declares to represent 'the Scylla and Charybdis of the problems of stingy exclusivity on one hand, and promiscuous inclusivity on the other'.[36] Crude definitions of religion in terms of belief in gods or spiritual beings—underlying Daniel Dennett's bold yet seriously inaccurate declaration that 'a religion without God or gods is like a vertebrate without a backbone'[37]—are rendered problematic by Buddhism, which obstinately refuses to conform to such definitions.[38] Cultural definitions of religion which focus on its outcomes—such as Clifford Geertz's view that religion is a system of symbols which evokes or establishes 'powerful, pervasive, and long-lasting moods and motivations'[39]—are rendered problematic by fanatical football followers and online gamers, who often display those same outcomes yet without conforming to conventional definitions of religiosity.[40]

These problems can, of course, be circumvented by suggesting that Buddhism is not really a religion, or that fanatical football fans are actually religious (without knowing it). Yet such evasions are ultimately seen as implausible, suggesting that the issue is not primarily about finding an improved definition of religion, but in fact that religion is not something that can be meaningfully defined scientifically in the first place, despite the perceived utility of arriving at some shared social understanding of what roughly we mean by the word.[41]

---

[36] Jong, 'On (Not) Defining (Non)Religion', 18.

[37] Dennett, *Breaking the Spell*, 9. Dennett's view of religion is clearly an uncritical reflection of his American cultural context: see Schaefer, 'Blessed, Precious Mistakes', 78 n. 6: 'Dennett's failure to pursue even a single secondary source on the "authorities" [for the definitions of religion] he cites suggests a problematic lapse of academic due diligence, one that is made possible by trafficking in a set of common-sense assumptions about religion that circulate in the American context.'

[38] Southwold, 'Buddhism and the Definition of Religion'. Southwold suggests that scholars tend to treat 'religion' as 'the polythetic class of all cultural systems that it seems reasonable to call religions'.

[39] Geertz, *The Interpretation of Cultures*, 90.

[40] The difference on this point is a matter of degree and not of kind: cf. Geraci, *Virtually Sacred*, 63–169, who emphasizes the role of mythology in many such games.

[41] See the points made by Harrison, 'The Pragmatics of Defining Religion in a Multi-Cultural World'; Schaffalitzky de Muckadell, 'On Essentialism and Real Definitions of Religion'.

Religion too easily becomes a definitional victim in a culture war—usually with the natural sciences.

As Peter Harrison's revisionist account of the complex historical relationship of science and religion makes clear, this can be (and has been) mapped in multiple manners, with significantly different outcomes. For example, consider the influential map developed by the social anthropologist J. G. Frazer in his *The Golden Bough* (1890), which depicts religion as a primitive form of science, offering explanations of the world which are refuted by modern science.[42] As Marilynne Robinson points out, this map of the territory of religion conceives it as 'a crude explanatory system, an attempt to do what science actually could do, that is, account for the origins and the workings of things'.[43] On the basis of this deficient understanding, science and religion must compete for the same logical space, and are to be judged by the same empirical criteria. Yet this negative evaluation of religion is dependent upon the reliability of Frazer's widely ridiculed conceptual map,[44] which maps religion in a specific—and ultimately indefensible—way.

Now compare this with a Wittgensteinian map of the territories of science and religion, which does not frame religion as a flawed version of science which is deficient in evidential foundations, sophistication, or predictive power, but rather as constituting certain shared practices, observances, and rituals that have a special significance to its practitioners. Such a map suggests that there is, and can be, no meaningful opposition between science and religion, in that they do not occupy the same logical space.[45]

Such a remapping of intellectual and imaginative possibilities—to which others might easily be added—creates ample conceptual space for exploring alternative readings of the relationship of science and

---

[42] Josephson-Storm, *The Myth of Disenchantment*, 125–52.

[43] Robinson, *What Are We Doing Here?*, 257.

[44] The work is now seen by anthropologists as 'a cautionary tale of a grand project blighted by its poor anthropological theory and methodology': see Kumar, 'To Walk Alongside'. Curiously, Richard Dawkins makes Frazer's discredited 'general principles' of religion central to his evolutionary debunking arguments: Dawkins, *The God Delusion*, 188.

[45] For a popular presentation of this kind of map, see Labron, *Science and Religion in Wittgenstein's Fly-Bottle*.

religion. Many now argue that the outdated traditional Western 'narrative of conflict' between science and religion needs to be replaced with a 'narrative of enrichment', which allows for the possible integration of scientific functionality with existential concerns, such as meaning and value.[46]

Yet in suggesting that the richness and complexity of the world should be reflected in the diversity and sophistication of our intellectual and imaginative reflections, the question of the rationality of such multiple approaches cannot be evaded. Does the study of the functioning of our universe entail standards of rationality which are divergent from—or are even in conflict with—those that we might use to explore questions of 'value and purpose'? Can what Dewey believed to be the 'deepest problem of modern life' be resolved without a lapse in or compromise of rationality? Is transdisciplinarity itself an incoherent enterprise, in that it lacks a *single* operational rationality? Or might we think in terms of distinct 'rationalities', each appropriate to its own domain of investigation, which have to be held together in a creative tension within individual minds? Or can we develop a richer or deeper concept of rationality, capable of extending across disciplines, finding a distinct implementation within each such discipline?[47] Or perhaps retrieving the notion of wisdom as the integration of disciplinary insights, and their incorporation into the world of life?[48] The exploration of such questions is an integral aspect of the agenda of this book.

THE AIM OF THIS BOOK

Building on some existing studies in this field,[49] this book aims to establish the understandings of rationality as both theory and practice

---

[46] e.g. McGrath, *Enriching our Vision of Reality*.
[47] For example, see Guattari, 'La transversalité'; Ryan, 'Wisdom, Knowledge and Rationality'.
[48] This is often seen as a correction of modernity's tendency towards disciplinary fragmentation: see, for example, Kaufman, 'Knowledge, Wisdom, and the Philosopher'.
[49] Major contributions to this field in the last few decades include Clayton, *Explanation from Physics to Theology*; Banner, *The Justification of Science and the*

encountered within professional communities in both the natural sciences and Christian theology, and to avoid simplistic reductions to allegedly 'essential' or 'universal' characterizations of either 'science' or 'religion', or the rational practices or criteria which emerge within them. The approach adopted in this work is fundamentally (though not exclusively) *empirical*, in that it sets out to consider how rationality is understood and enacted, rather than to offer predetermined accounts of what forms rationality ought to take. The patterns of rationality that such an undertaking discloses are complex, and resist easy reduction to categories and taxonomies. The present study thus stands at some distance from those more abstractly philosophical approaches to rationality which are disinclined to consider how rationality is actually understood within specific epistemic communities, and implemented in their virtues and practices.

This book thus sets out to map the concepts of rationality and their attendant practices across disciplinary fields, with a particular focus on the fields of science and religion. To make this project manageable, it has proved necessary to restrict its focus to the relation of the natural sciences and Christian theology. Although this work engages extensively with the philosophy of science—for example, when considering the nature of explanation, or criteria of theory choice—its approach sometimes leads it to focus on specifically Christian themes, perhaps most notably an explicitly Trinitarian view of God (as opposed to generalized forms of theism).

The exploration of the relation of science and theology offers a case study in both interdisciplinarity and transdisciplinarity, which many now regard as a necessary response to the growing fragmentation of knowledge, the increasing disconnection between the academy and wider culture, and a wider concern to grasp a bigger picture of reality than one intellectual standpoint or academic discipline can offer. Such concerns lie behind recent integrative projects, such as Roy Bhaskar's critical realism, Ken Wilber's integral theory, Edgar Morin's *pensée complexe*, and E. O. Wilson's concept of consilience.[50] Yet any such

---

Rationality of Religious Belief; Stenmark, *Rationality in Science, Religion, and Everyday Life*; van Huyssteen, *The Shaping of Rationality*.

[50] For a summary and critical assessment of such approaches, see Marshall, 'Towards a Complex Critical Realism'; Segerstrale, 'Wilson and the Unification of Science'.

undertaking has to come to terms with the notion of 'multiple situated rationalities', and its implications for any form of dialogue or synthesis. In my view, this has not yet taken place; this book aims to establish a framework which is dependent neither on an outdated foundationalism nor on its antithetically conceived alternatives.

A hankering after simplification can easily distort our understanding of complex phenomena[51]—such as any informed attempt to understand human rationality which takes the complexity of its cultural and disciplinary manifestations seriously. Although the social-scientific stereotyping of religious thought as 'primitive' is now falling out of fashion as the ideological freighting of such approaches becomes increasingly obvious,[52] there still remains a reluctance in some circles to take the rationality of religious systems seriously. Yet it is clear that most religious people act according to what they regard as rational principles, which they consider to be justified and reasonable.[53] It is therefore important to attempt to explore how these map on to broader concepts of rationality, not least given the recognition of the multiple forms that human rationality takes.

It remains for me to thank colleagues who have stimulated my own thinking in this field, and above all in relation to the complex relationship between scientific and theological rationality. This book has been in preparation for many years, and I owe especial thanks to Charles A. Coulson (1910–74), Jeremy R. Knowles (1935–2008), Thomas F. Torrance (1913–2007), and R. J. P. Williams (1926–2015), whose influence can be discerned at many points within these pages, even if it is not always explicitly referenced. I also wish to acknowledge helpful conversations with many others, especially John Hedley Brooke, Joanna Collicutt, Peter Harrison, David Livingstone, Andrew Pinsent, Donovan Schaefer, Graham Ward, and Johannes Zachhuber. They cannot be blamed for the imperfections of this work, for which I take full responsibility.

Alister McGrath

Oxford University
May 2018

---

[51] Dupré, 'The Lure of the Simplistic'.
[52] See especially Iannaccone, Stark, and Finke, 'Rationality and the "Religious Mind"'.
[53] Witham, *Marketplace of the Gods*, 17–32.

I

Exploring the Notion of Rationality

I

Exploring the Notion of Rationality

1

One Reason; Multiple Rationalities

The New Context of Discussion

Je pense, donc je suis.[1] Descartes's maxim of 1637—usually cited in its later Latin form, *cogito ergo sum*—was as much an affirmation of the characteristic human capacity to think as it was a proposal for finding a secure foundation for knowledge in the face of the radical scepticism of his age.[2] The human propensity to think is often framed in terms of 'reason', which John Locke defined as 'a *faculty* in man, that faculty whereby man is supposed to be distinguished from beasts'.[3] Locke's caution about human uniqueness here is to be welcomed, given our growing interest in the phenomenon of animal reasoning.[4] Yet this does not in any way negate or subvert the decisive role of reasoning in human life and culture.

The unassailable empirical observation that human beings think cannot, however, be detached from more troubling normative questions about how they *should* think.[5] If we conceive rationality as 'a state or quality of being in accord with reason,' we need to be able to offer criteria by which we can judge the extent to which a specific

---

[1] Descartes, *Discours de la méthode*, 36–7. For Sartre's quite distinct reflections on the role of reason in relation to individual human identity, see Immel, 'Vom vernünftigen Ich'.
[2] Rott and Wagner, 'Das Ende vom Problem des methodischen Anfangs'.
[3] Locke, *Essay Concerning Human Understanding*, 667.
[4] For David Hume's reflections on this theme, see Boyle, 'Hume on Animal Reason'.
[5] Broome, 'Does Rationality Give Us Reasons?'; Kolodny, 'Why Be Rational?'; Reisner, 'Is there Reason to Be Theoretically Rational?'

belief, or process of argumentation, can be demonstrated to be 'in accord with reason'. Aristotle's categorial notion of reason is unhelpful here, in that its alternative is not so much *irrationality* as *arationality*—that is to say, acting in a manner that is not the outcome of reasoning.[6] Normative models of optimal judgement and decision-making—such as those developed by the empirical sciences—tend to understand rationality in terms of its proximity to the optimum defined by normative models of human cognition.[7] 'If human beings can indeed be described as rational animals, it is precisely in virtue of the fact that humans, of all the animals, are the only ones capable of irrational thoughts and actions.'[8] If a behaviour or belief is to be described as irrational, it must be capable of being shown that it departs from the optimum prescribed by a particular normative model.

So are there norms to which reasoning ought to conform? And if so, how might such norms be justified without the implicit presupposition of their validity?[9] As Wittgenstein pointed out, the complexities of human reasoning processes are such that one and the same proposition or idea may at one point be treated as something that is *to be tested*, and at another as a *rule of testing*.[10] The discussion of human rationality, particularly since the eighteenth century, has struggled with the tension between the descriptive and normative on the one hand, and the apparently irresolvable circularity of discussions of rationality on the other. These tensions are exposed, rather than resolved, by attempts to distinguish the 'reasonable' from the 'rational,'[11] and are often exacerbated by the rhetorical tendency to elevate

---

[6] De Sousa. *Why Think?*, 6–7.
[7] Chater and Oaksford, 'Normative Systems'.
[8] De Sousa. *Why Think?*, 7.
[9] The problem of rational 'bootstrapping', a circular form of argument in which reason itself is used to ground or evaluate reason, should be noted here. Bootstrapping is often defined using the statement 'a belief that an epistemic rule R is reliable can be justified by the application of R'. See Cheng-Guajardo, 'The Normative Requirement of Means-End Rationality and Modest Bootstrapping'. There is widespread hostility to any form of bootstrapping in the philosophy of science, even if this is not always accompanied by a precise account of how it is to be understood, and thus averted: see Vogel, 'Epistemic Bootstrapping'.
[10] Wittgenstein, *On Certainty*, 98. For further discussion, see Ariso, 'Unbegründeter Glaube bei Wittgenstein und Ortega y Gasset'.
[11] For significant contributions to this discussion, see Sibley, 'The Rational Versus the Reasonable'; Rawls, *Political Liberalism*, 48–53.

the status of the problematic notion of 'rationality' to that of a cultural virtue. Broad statements such as 'it is not rational to believe in God' or 'it is rational to believe in the goodness of human nature' often rest on unacknowledged normative cultural judgements, which clearly need to be identified and interrogated.

In this opening chapter, I shall offer some introductory reflections (perhaps best seen as 'orientating generalizations'[12]) which need to be considered before engaging in a more detailed assessment of the theme of rationality in science and religion.

SHIFTING NOTIONS OF RATIONALITY

Rationality is regularly affirmed as a cultural virtue,[13] a core element of the self-image of the human species.[14] Yet this admirable virtue proves remarkably resistant to clarification, often serving as a cultural aspiration which seems to lie beyond both a meaningful consensual definition on the one hand, and empirical verification on the other.[15] Wittgenstein's assertion that rationality has a history, and takes different forms in different social locations, resonates with a growing empirical and historical awareness of the diversity of human concepts of rationality. 'What men consider reasonable or unreasonable alters. At certain periods, men find reasonable what at other periods they found unreasonable. And vice versa.'[16]

Human cultures possess and exhibit multiple notions of rationality, concepts of sense-making, and criteria for adjudication of the truthfulness of beliefs.[17] A survey of the manner in which human rationality

---

[12] I borrow this phrase from Ken Wilber, and use it in his sense: Wilber, *Sex, Ecology, Spirituality*, 5.
[13] For Plato's political application of reason, see Miller, 'The Rule of Reason in Plato's *Laws*'.
[14] Nozick, *The Nature of Rationality*, xii.
[15] Nickerson, *Aspects of Rationality*, 1–70; Lenk, 'Typen und Systematik der Rationalität'.
[16] Wittgenstein, *On Certainty*, 336.
[17] Moore, 'Varieties of Sense-Making'. More generally, see the important collection of essays in Apel and Kettner, eds, *Die eine Vernunft und die vielen Rationalitäten*. For a similar approach, critiquing normative western assumptions about rationality,

has been understood and represented from the seventeenth to the late twentieth centuries discloses radical shifts in patterns of thinking,[18] suggesting that rationality is not a given universal norm, but is embedded within its own *Lebensform* and *Lebenswelt*.[19] The idealization of human rationality by some Enlightenment thinkers led to its being seen as culturally and historically invariant, capable as serving as both the foundation and the criterion of reliable knowledge.[20] Antoine Destutt de Tracy thus introduced the notion of 'ideology' in post-revolutionary France to refer to the domain of ideas which he believed arose unproblematically from the exercise of human rationality.[21] Yet this assumption, while once seen as culturally plausible and intellectually generative, is now generally seen as a period piece, reflecting assumptions which it is now difficult to endorse.

There has been a growing realization that both the beliefs that we hold and the rationality through which we develop and assess these beliefs are embedded in cultural contexts. Rationality is thus increasingly coming to be seen as being dependent (though questions remain about the extent and nature of that dependency) upon its historical and cultural context, and best assessed in terms of the practices it generates.

RATIONALITY, EMBODIMENT, AND EMBEDDEDNESS

Over the past twenty-five years, several claims about human cognition and its underpinnings have gained traction within the academic

---

see the collection of essays reflecting an African perspective in Hountondji, ed., *La rationalité, une ou plurielle?*

[18] See the comprehensive study of Wollgart, 'Zum Wandel von Rationalitätsvorstellungen vom 17. bis zum 20. Jahrhundert'.

[19] Abel, 'Der interne Zusammenhang von Sprache, Kommunikation, Lebenswelt und Wissenschaft'. There are some important parallels here with the approach of the Spanish philosopher José Ortega y Gasset, especially his essay 'Ideas y creencias', with its landmark statement 'Las ideas se tienen; en las creencias se está'. See further Defez, 'Ortega y Wittgenstein'.

[20] Flyvbjerg, *Rationality and Power*, 1–8.

[21] These ideas were consolidated and systematized in his *Eléments d'idéologie* (1801–15).

research community. It is now clear that human beings regularly and characteristically reach decisions on the basis of what can be shown to be flawed reasoning processes. Recent discussions in the philosophy of knowledge have highlighted our epistemic fallibility, in that our psychological commitment to our views often turn out to significantly exceed the actual epistemic justification for those views.[22] It is now also clear that cognition is affective—that is, that it is intimately connected with the perceived value of the object of cognition to the observer.[23] This recognition of the role of emotions helps us understand the deeper rationality of seemingly irrational decisions or 'akratic' actions.[24] Yet particular attention has focussed on the recognition that human cognition is both *embodied* and *embedded*. Our cognitive properties and performances are often significantly dependent on our embodiment, and our relationship to our physical and cultural environment.[25]

The starting point for any twenty-first-century discussion of rationality—whether as a general human phenomenon or as a specific factor in understanding how specialist communities develop and validate their ideas—is thus a recognition that human beings are both *physically embodied* and *culturally embedded*.[26] This recognition of the historicity of rationality avoids the forms of ahistorical thinking characteristic of so many forms of rationalism, and insists that history should be allowed to disclose norms of rational choice. The justification of such choices does not result from human reason as such, but is rather grounded in the particular communities in which we find ourselves located.[27] Whereas older writers regarded a culturally and

---

[22] Fumerton, 'Why You Can't Trust a Philosopher'; Leite, 'Believing One's Reasons Are Good'; Hammond, *Beyond Rationality*, 145–90.

[23] Columbetti, 'Enaction, Sense-Making and Emotion'. For the importance of this point to theology, see Zahl, 'On the Affective Salience of Doctrines'.

[24] Akratic actions can be defined as free and intentional actions performed despite the judgement that another course of action is better. For discussion of this point, see Tappolet, 'Emotions and the Intelligibility of Akratic Action'.

[25] Clark, *Being There*; Haugeland, *Having Thought*; Gallagher, *How the Body Shapes the Mind*.

[26] For the importance of this point from anthropological and psychological perspectives, see Marchand, ed., *Making Knowledge*; Shapiro, *Embodied Cognition*.

[27] Note especially John Stuart Mill's notion of socially embedded rationality: Zouboulakis, *The Varieties of Economic Rationality*, 14–24.

historically invariant reason as the prime source of authority for their social and political views,[28] this view came to be seen as philosophically misguided and politically dangerous by many liberal writers, in that it assumed an ahistorical definition or criterion of 'reasonableness' which led to the marginalization or social devaluation of allegedly 'unreasonable' views.[29]

The way in which human beings sense, perceive, and interpret information about the world around them is now agreed to be partly determined by both psychological and social factors.[30] Natural human cognitive processes are contextualized within, and modulated by, a sociocultural environment. The insight that human beings are culturally embodied means that the manner in which we sense, perceive, and interpret the world around us is determined by both psychological and social influences. This point is integral to the present study, which distances itself from more abstractly philosophical approaches to rationality, characteristic of some earlier discussions of rationality in science and religion, which fail to ascertain how rationality is actually understood in practice, and how this can be explored and interpreted empirically. It is therefore important to identify the embedded cultural assumptions which shape a culture's understandings of rationality or interpretations of the natural world,[31] partly to subvert any claims of ultimacy for any given way of understanding such notions, but more fundamentally to understand the critical role of a specific cultural context in interpreting human experience.[32] A belief might seem to be rational in terms

---

[28] Galston, 'Two Concepts of Liberalism', 525.

[29] Mouffe, *The Return of the Political*, 141–4. For Mouffe, this recognition of a historicized rationality inevitably leads to an 'agonistic pluralism': Mouffe, *The Return of the Political*, 1–8.

[30] See Frank, 'Sociocultural Situatedness'; Kimmel, 'Properties of Cultural Embodiment', especially 91–5.

[31] For a detailed analysis, see Avrahami, *The Senses of Scripture*, 4–64.

[32] The term *sensorium* is now sometimes used to designate the distinct and shifting sensory environments within which individuals are located, determined by both natural capacities and cultural influences, which shape how we understand ourselves and our world. See Ong, 'The Shifting Sensorium'. For its theological application, see McGrath, *Re-Imagining Nature*, 41–61. This modern use of the term differs from that of Isaac Newton, who conceived it primarily as the human sensory apparatus: Kassler, *Newton's Sensorium*, 3–27.

of the criteria embedded in one sociocultural situation, but not in another.

Three elements can be identified in shaping the perceptions of what beliefs are rational, although the manner of their interaction is the subject of continuing research and reflection:

1. Natural human cognitive processes, which are generally thought to be independent of a given individual's cultural location;
2. The prevailing cultural metanarrative, which influences our judgements about what is to be deemed 'reasonable' or corresponding to 'common sense';
3. The evidence that is available to individuals in that specific cultural location.

Each of these three elements clearly requires further discussion.

1. *Natural human cognitive processes.* The way in which human beings assess situations and arrive at 'rational' conclusions has been the topic of intense empirical study since the early 1970s. The heuristics and biases research programme inaugurated by Kahneman and Tversky clearly demonstrated that descriptive accounts of human behaviour diverged from normative models.[33] Since then, it has become increasingly clear that supposedly rational individuals are prone to assess probabilities incorrectly, test hypotheses inefficiently, fail to calibrate degrees of belief properly, ignore alternative hypotheses when evaluating data, as well as displaying other significant information-processing biases.[34] These defects can, at least to some extent, be corrected, once the nature of the problem is grasped.

2. *The prevailing cultural metanarrative.* Social theorists such as Cornelius Castoriadis and Charles Taylor have explored how the 'social imaginaries' of any specific cultural location play a significant role in what is deemed to be 'rational' or to constitute 'common sense'.

---

[33] For example, Kahneman and Tversky, 'Subjective Probability'. The key point is that humans find it difficult to make probability judgements. Kahneman won the Nobel Prize for Economics in 2001.

[34] For a review of the literature, see Samuels and Stich, 'Rationality and Psychology'; Stanovitch, 'On the Distinction between Rationality and Intelligence'.

Taylor suggests that our cultural imagination has now been captured by a 'picture' that leads us to deem certain ideas, which might once have seemed to be eminently 'reasonable', irrational. As we have already noted, a belief that was thought to be rational in one context is thus considered irrational in another. The belief itself may not have changed; the context within which it is evaluated, however, has altered, with significant implications for the cultural plausibility of that belief. Given its importance for understanding the cultural enactment of human rationality, we shall consider this point in more detail later in this work (84–6).

3. *The available evidence.* Humans reflect on the evidence that is available to them. Yet such evidence is often determined by the contingencies of history and culture. For example, the observation of galactic spectral red shifts, which pointed to an expanding universe and hence to a moment of cosmic origination, was simply not possible before the First World War. Or, to give another example, the relatively late development of the notion of biological evolution is not an indication of flawed human reasoning in earlier periods, but of the absence of evidence which made such ideas initially plausible, and subsequently compelling.[35] Reasoning about our world is inevitably limited by what we can see of that world.

These three broad categories of factors help us to organize the various contributories to the flux of human rationality, offering an interpretative framework that goes some way towards accounting for variations in patterns of human thinking across history and culture. It avoids reducing questions of rationality to the proper functioning of human cognitive processes, while nevertheless recognizing that such processes are of critical importance. It recognizes the cultural embeddedness of individual thinkers and interactive groups of thinkers, while making allowance for the influence of their specific historical location on the intellectual resources at their disposal—such as observational evidence. Human thinkers are embodied, existing in a complex relationship with their physical and social environment, involving both top-down and bottom-up interactions which make it impossible to treat cognitive functioning in a culturally or socially

---

[35] Corsi, *Evolution before Darwin.*

detached manner.³⁶ The human mind creates culture, which in turn interacts with the manner in which that mind functions, thus creating a complex layered framework of interaction and feedback.

A failure—however understandable—to appreciate that criteria of what is 'reasonable' are culturally situated helps us to make sense of the otherwise puzzling failure of the English Enlightenment to recognize that the notion of a 'universal human rationality' was called into question by the 'voyager' literature of the seventeenth and early eighteenth centuries, which reported on cultures whose notions of morality and rationality diverged significantly from those of England.³⁷ Instead of allowing such observations to challenge the emerging myth of a universal rationality, leading to a remapping of the territories of human reason, the English Enlightenment generally tended to prefer creating social binaries, such as 'civilization—savagery' and 'rational—irrational', to accommodate (and thus to *neutralize*) these anomalies within the framework of this totalizing and universal notion of reason.³⁸ These observations raise concerns about the ethnocentric approaches to discussions of rationality often regarded as characteristic of the European Enlightenment, which suggest to more critical observers that thinkers embedded within this Western philosophical tradition had made themselves 'prisoners of their own sensorium'.³⁹

REFLECTIONS ON THE CULTURAL AND SOCIAL EMBEDDEDNESS OF RATIONALITY

The cultural specificity of notions of rationality is reflected in the shifting cultural assessment of the rationality of certain religious notions—such as belief in God, or in special divine action—in western Europe and North America. The notion of divine action was predominantly seen as

---

³⁶ Shea, 'Distinguishing Top-Down from Bottom-Up Effects'. See further Ellis, *How Can Physics Underlie the Mind?*, especially 372–6.
³⁷ Carey, *Locke, Shaftesbury, and Hutcheson*, 14–97.
³⁸ Gozdecka, *Rights, Religious Pluralism and the Recognition of Difference*, 139–44.
³⁹ Avrahami, *The Senses of Scripture*, 10.

entirely reasonable in the sixteenth century, yet is generally viewed as requiring extended justification in the twenty-first.[40] Charles Taylor highlights this shifting cultural perception of rationality by focussing on the perceived reasonableness of belief in God. 'Why was it virtually impossible not to believe in God in, say, 1500 in our Western society, while in 2000 many of us find this not only easy, but even inescapable?'[41]

Taylor's answer to this shifting perception of what counts as 'rational' highlights the importance of culturally regnant metanarratives in shaping a society's understanding of what is 'reasonable'.[42] The wisdom of Taylor's approach can be appreciated by considering scholarly reconstructions of the worldviews of educated Europeans around 1600—with such features as seeing comets as signs of evil, believing in the possibility of changing base metals into gold, and acknowledging an influence of the planets on human destiny.[43] Perceptions of what is 'rational' are, at least in part, socially constructed, in that these are influenced, if not determined, by a set of historical contingencies, such as the character of the dominant cultural metanarrative. For Taylor, to hold patterns of thought that run counter to a deeply embedded cultural mindset or groupthink is potentially to be seen and judged as irrational. This judgement arises through the exercise, not of pure reason, but of culturally shaped notions of what counts as 'rational' within a culture or community of discourse, highlighting the need to recognize the social dimensions of rationality.

A similar point needs to be made in relation to the notion of 'common sense', which is often presented, especially in popular discussions of issues of rationality, as a straightforward and universal account of the way things are—or ought to be. Nicholas Rescher, for example, speaks of common sense as 'the commonplaces of everyday-life experiences of ordinary people in the ordinary course of things'.[44] Yet just who are these 'people'? Who decides what is 'ordinary'?

---

[40] For the issues, see McGrath, 'Hesitations about Special Divine Action'.

[41] Taylor, *A Secular Age*, 25.

[42] Taylor highlights the importance of the emergence of what he terms an 'immanent frame' and 'closed world structures' within this process: Taylor, 'Geschlossene Weltstrukture in der Moderne'.

[43] Wootton, *The Invention of Science*, 6–7.

[44] Rescher, *Common Sense*, 11.

And who interprets these 'experiences'? For Heidegger, common sense is part of the public interpretation of the world, an aggregate of 'self-evident' truisms that characterizes a group's normative self-conception.[45] As Clifford Geertz observed, the 'unspoken premise from which common sense draws its authority' is that it 'presents reality neat'.[46] Yet closer examination indicates a need for the relocation of this authority within a broader cultural context. 'If common sense is as much an interpretation of the immediacies of experience, a gloss on them, as are myth, painting, epistemology, or whatever, then it is, like them, historically constructed and, like them, subjected to historically defined standards of judgement.'[47]

Geertz is particularly critical of anthropologists who bring 'commonsense' reasoning to bear on the social habits and worlds of thought of those whom they find culturally alien—such as Evans-Pritchard's famous discussion of Azande witchcraft in his influential *Oracles, Witchcraft, and Magic Among the Azande* (1937). Evans-Pritchard may have thought he was a neutral objective observer; in fact, he was a culturally embedded observer, who judged another cultural system on the basis of his own beliefs, which were assumed to be correct. Geertz argued that what Evans-Pritchard regarded as a crude and primitive metaphysical system simply represented Azande commonsense assumptions.[48] Indeed, recent research has suggested that many cultures merge a range of causal explanations of events, and do not see this as entailing contradiction or tension.[49]

This general question of the cultural embeddedness of a rational observer was explored earlier in the great 'rationality debate' within the philosophy of social sciences, which focussed particularly on the difficulties faced by Western observers in interpreting and assessing the rationality of an alien culture from the rational perspectives of their own.[50] Peter Winch suggested that Evans-Pritchard was

---

[45] Absher, 'Speaking of Being', 210–12.
[46] Geertz, 'Common Sense as a Cultural System', 8.
[47] Geertz, 'Common Sense as a Cultural System', 8.
[48] Geertz, 'Common Sense as a Cultural System', 10–12.
[49] See Legare, Evans, Rosengren, and Harris, 'The Coexistence of Natural and Supernatural Explanations across Cultures and Development'.
[50] Davies, 'Living in the "Space of Reasons"'.

judging the beliefs of a primitive society on the basis of his own (unacknowledged) beliefs. Winch's essay explored the difficulty in attempting 'to make intelligible in our terms institutions belonging to a primitive culture, whose standards of rationality and intelligibility are apparently quite at odds with our own.'[51]

Winch's questions may not have been answered to everyone's satisfaction; the issues he raises, however, remain significant. One obvious point concerns the failure of secular writers to appreciate how theological ideas can inform and stimulate the intellectual and moral vision of individuals and communities; since they do not share these views, they cannot grasp the explanatory power of theology as an intellectual system of epistemic weight.[52] The historian Brad Gregory has drawn attention to the 'uncritically accepted presuppositions' that prevent secular historians from appreciating the descriptive and illuminating function that theological frameworks had in shaping the mentality of early modern religious individuals.[53] In effect, such writers are privileging their own perspectives, and in doing so lack the intellectual empathy necessary for the social scientific virtue of *Verstehen*. It is becoming clear that much of the received modernization model is not so much a scientific description of actual processes of social change, but rather a structure resting on a specific normative antireligious ideology of 'progress' and a secular conception of a 'good society', presented as if they were objective social science.

There is another point that needs to be made here. Nineteenth-century writers tended to assume that culture was an invariant, something absolute that did not vary from one location to another, save in the extent to which a given society was 'cultured'. After the First World War, perceptions changed. No longer was it realistic to speak of 'culture' in the singular; it was finally recognized that there existed (and presumably always have existed) a range of cultures, each of which was bounded, coherent, cohesive, and self-standing.[54] Perhaps most importantly for our purposes, each culture came to

---

[51] Winch, 'Understanding a Primitive Society'. For comment, see Schilbrack, 'Rationality, Relativism, and Religion'.
[52] See the rich collection of material in Chapman, Coffey, and Gregory, *Seeing Things Their Way*.
[53] See the analysis in Gregory, 'No Room for God?'
[54] Galison, 'Scientific Cultures', 121–2.

be seen as having developed its own understanding of what was 'reasonable'. This has raised important questions for normative judgements about the rationality of certain beliefs and practices—such as those traditionally described as 'magic'.[55]

If Thomas Nagel is correct in suggesting that we cannot escape the condition of seeing the world from our particular spatial and cultural insertion within it, our concepts of rationality will be shaped to some extent and in some manner by the traditions of discourse and research within which we find ourselves located.[56] Academics are not exempt from the contingencies of cultural and historical location. Most academic disciples experience significant shifts in their self-understandings and underlying values over time, leading to changes in dominant paradigms and the scholarly consensus. Such concepts as 'objectivity' and 'impartiality' attained the status of epistemological virtues in the late seventeenth and eighteenth centuries, largely because they bolstered the plausibility of a 'view from nowhere', untainted by partisan concerns or cultural precommitments.[57] Yet the fact that these two notions have a history—that is to say, that they emerged as significant at a particular moment in history—points to the historical particularity of the concepts of rationality associated with the 'Age of Reason'.

Similar trends can be seen emerging within recent empirical studies of human logic. Whereas 'logic' was once thought to imply what we might now prudently designate as 'classical logic', it is now becoming clear that multiple 'logics' develop to meet specific tasks in knowledge production. This trend has spread from within the somewhat restricted mathematical study of logic to the use of logic in normative theories of everyday human activities, so that a plurality of logics emerges in which the context of the agent's goals determines the relevant logical norms.[58]

---

[55] For a careful exploration of this concern from an anthropological and sociological perspective, see Sanchez, *La rationalité des croyances magiques*.

[56] Nagel, *The View from Nowhere*, 67–89.

[57] As argued by Daston, 'Objectivity and the Escape from Perspective'; Murphy and Traninger, eds, *The Emergence of Impartiality*.

[58] Achourioti, Fugard, and Stenning, 'The Empirical Study of Norms Is Just What We Are Missing'. See further Beall and Restall, *Logical Pluralism*.

THE EMBODIMENT OF RIGHT REASON: THE 'WISE'

Reason is an embodied activity. For this reason, Aristotle's dialectical method often involves an appeal to *ta endoxa*—'those things which are accepted by everyone, or by most people, or by the wise (*sophoi*)'.[59] Aristotle's category of *ta endoxa* remains somewhat opaque, but clearly designates those beliefs which are accepted by sociologically determinative or privileged groups—above all, the 'wise'.[60] Although Aristotle's main concern in developing his endoxic method is to develop a perspective which can both explain and support these 'common beliefs' or bring them into harmony with one another, it is impossible to overlook his assumption that the views of certain social groups are deemed to be privileged.

Similar difficulties in defining 'reason' as an operational concept became evident in the early modern period. In a world that increasingly considered individual reason and private judgement to be the primary arbiters of truth, there was clearly a need to be able to navigate the stormy sea of conflicting reality claims. For its critics, the grand narrative of Enlightenment could not be disentangled from Eurocentric biases and power dynamics, and was stubbornly resistant to any form of epistemic reflexivity.[61]

One way of dealing with this problem emerged in the middle of the seventeenth century—to create a social hierarchy where some voices are valued more than others because they are held to personify 'sound reason' or embody 'right thinking'. Thomas Hobbes, finding himself forced to the conclusion that there was no self-evidently correct mode of reasoning embedded within the natural order, suggested that this 'want of a right Reason constituted by Nature' required the reinterpretation of 'right Reason' in terms of 'the Reason of some Arbitrator,

---

[59] Aristotle, *Topics* I.1; 100a18-2; *Nicomachean Ethics*, VII.1; 1145b4.

[60] For a detailed assessment, see Bolton, 'The Epistemological Basis of Aristotle's Dialectic'. Cf. Smith, 'Logic', 60–1.

[61] These same criticisms can be made against more recent attempts to develop metatheories, such as Integral Theory: see Rutzou, who argues in 'Integral Theory and the Search for the Holy Grail' that it is 'quintessentially western and illicitly universalizing'.

or Judge', to be agreed by those wishing to resolve their debates.[62] Since there is no such thing as 'right reason', the reason of some individual or group of individuals must serve as a proxy. For Hobbes, those who appealed to 'right reason' thus generally had their own ideas in mind. 'This common measure, some say, is right reason: with whom I should consent, if there were any such thing to be found or known *in rerum naturâ*. But commonly they that call for *right reason* to decide any controversy, do mean their own.'[63]

This tendency to see 'right reason' or 'right thinking' embodied or exemplified in luminous exemplars is, as Hobbes rightly discerned, ultimately a matter of social judgement or convention. The emergence of a notional 'reasonable person' who served both as an instantiation and criterion of 'right thinking' during the eighteenth century reflects this difficulty in defining normative modes of operative reasoning, and reveals how covert sociological norms or influences often intruded into debates which were notionally about a disembodied universal rationality. 'Right-thinking people' came to designate the cultural icons of social in-groups in England, France, and Germany, so that debates about rationality all too easily became transposed into issues of social alignment rather than of epistemic virtue.[64]

In the last twenty years, there has been intense scholarly scrutiny of the defining characteristics of the Enlightenment. Many have framed this movement using Kant's famous essay 'Was ist Aufklärung?' (1784), which has become for many a 'one-stop shop for defining the Enlightenment'.[65] For Kant, 'Enlightenment (*Aufklärung*) is the emergence of humans from their minority (*Unmündigkeit*), for which they themselves are to blame. Minority is the inability to use one's reason without the guidance of another.'[66] This is probably best seen as a retrospective attempt to impose coherence on a complex cultural movement, with significant regional variations and agendas, through the articulation of a distinctive narrative or conceptual framework

---

[62] Hobbes, *Leviathan*; *English Works*, vol. 3, 31.
[63] Hobbes, *De Corpore*; *English Works*, vol. 4, 225.
[64] Freeden, *Ideologies and Political Theory*, 31–2.
[65] Edelstein, *The Enlightenment*, 117.
[66] Kant, *Was ist Aufklärung?*, 5. For the modification of Kant's definition in the later German Enlightenment, see Gödel, 'Eine unendliche Menge dunkeler Vorstellungen'.

that was flexible enough to allow the diverse elements of contemporary intellectual culture to have at least some sense of shared purpose and identity.

Yet we must note that there are multiple narratives of identity of this kind, originating from different quarters of the Enlightenment, and it is difficult to see who has the authority to determine which is *the* defining narrative.[67] Kant's definition of the Enlightenment may be the most-cited today; it is, however, a summative statement framed with a definite agenda in mind.

There is no doubt of the significance of the Enlightenment appeal to human reason, partly as a means of avoiding religious debates about the interpretation of sacred texts. Yet some suggest that the Enlightenment is best understood not as 'an aggregate of ideas, actions, and events', but rather as a narrative which 'provided a matrix in which ideas, actions, and events acquired new meaning'.[68] On this reading, the human use of reason was thus located within a new narrative framework. Yet debates about the limits and proper use of reason highlighted the need to calibrate this new confidence in human rationality.

A good example of this kind of discussion is found in Jean-Baptiste le Rond d'Alembert's 'Discours préliminaire' (1751) in the celebrated *Encyclopédie*, which declared that 'the art of reasoning is a gift which Nature bestows of her own accord upon people of discernment (*bons esprits*)'.[69] The category of 'reason', which earlier generations had grounded in theological principles, was now seen as embodied in cultural exemplars, who played a critical role in the establishment of a 'rational' authority. As Hobbes had foreseen, the virtue of being 'rational' thus became a matter of intellectual and cultural

---

[67] See, for example, Schmidt, 'Inventing the Enlightenment'; Hunt and Jacob, 'Enlightenment Studies'.

[68] Edelstein, *Enlightenment*, 13.

[69] D'Alembert, 'Discours préliminaire'; *Oeuvres*, vol. 1, 33. 'L'art de raisonner est un présent que la Nature fait d'elle-même aux bons esprits'. D'Alembert regularly uses the term 'bons esprits' to refer to right-thinking individuals, seeing such an intellectual oligarchy as culturally determinative: see Hayes, *Reading the French Enlightenment*, 45. Historically, D'Alembert's *Discours* and Condorcet's *Esquisse d'un tableau historique des progrès de l'esprit humain* (1795) were especially influential in shaping the self-understanding of the Enlightenment.

alignment with approved exemplars, so subtly transposing questions of rationality into those of conformity with the views of privileged authorities. A generation later, Nicolas de Condorcet sought to avoid the problem of such cultural oligarchies by a rationalization of processes of judgement that in effect amounted to a mechanization or mathematization of decision-making,[70] which made the social location of those who made those decisions theoretically irrelevant, yet—as events demonstrated—practically decisive.

Nietzsche is one of many critics of the Enlightenment vision of rationality, arguing that human reason does not simply and neutrally render the way things are, but schematizes them, and reshapes them in the form of falsehoods and distortions, trying to force a complex reality into a simple intellectual framework. Nietzsche thus highlights the way in which we do violence to the singularity and variability of phenomena by means of generalizations and universalizations.[71]

CONCERNS ABOUT HUMAN RATIONALITY

The optimism of the Enlightenment partly reflected what now seems to be an unrealistic estimate of the capacity of human reason on the one hand, and the benign nature of humanity on the other. The experience of destructive global wars in the twentieth century seemed to call both into question. As R. G. Collingwood gloomily remarked on the eve of the Second World War, some 'the chief business of twentieth-century philosophy is to reckon with twentieth-century history'.[72] Echoing such concerns, Bertrand Russell argued that human rationality was a noble aspiration that unfortunately lagged far behind the reality disclosed by the harsh realities of observation.

Writing in the midst of the Second World War, Russell found himself overwhelmed with what seemed to be incorrigible evidence

---

[70] Bates, *Enlightenment Aberrations*, 78–85; Kavanagh, 'Chance and Probability in the Enlightenment'.
[71] Welsch, 'Nietzsche über Vernunft'.
[72] Collingwood, *An Autobiography*, 79. For the wider context of such concerns, see Glover, *Humanity*.

of human irrationality, which made him despair about the human future. 'Man is a rational animal—so at least I have been told. Throughout a long life, I have looked diligently for evidence in favor of this statement, but so far I have not had the good fortune to come across it.'[73] The subsequent emergence of 'Cold War rationality' in the late 1950s, based on algorithms of 'mutually assured destruction', only seemed to confirm Russell's dark anxieties.[74] Many would now argue that such seemingly irrational outcomes reflect the 'collectivizing of human reason', in which a group collectively endorses a conclusion that most of the group would individually reject.[75] This, however, merely complexifies the problem of human rationality, without resolving the representative concerns raised by Russell.

For Russell, human aspirations to rationality were compromised by the destructive 'intellectual vice'[76] of a natural human craving for certainty, which could not be reconciled with the capacities of human reason on the one hand, and the complexity of the world on the other. Philosophy, Russell suggested, was a discipline deeply attuned to this dilemma, enabling reflective human beings to cope with their situation. 'To teach how to live without certainty, and yet without being paralyzed by hesitation, is perhaps the chief thing that philosophy, in our age, can still do for those who study it.'[77]

Many, of course, would disagree with Russell, whose own epistemological agnosticism is often overlooked by those who simplistically classify him as an atheist.[78] The resurgence of religious fundamentalism on the one hand, and popular forms of scientific positivism on the

---

[73] Russell, *Unpopular Essays*, 82.

[74] Russell, *Common Sense and Nuclear Warfare*, 30. Cf. Erickson et al., *How Reason Almost Lost Its Mind*, especially 85–7. Russell's concern was that essentially irrational choices would be made under pressure, leading to nuclear devastation: cf. Quackenbush, *International Conflict*, 179–86.

[75] Pettit, 'Groups with Minds of Their Own', especially 175–8. More generally, see Rupert, 'Minding One's Cognitive Systems'. The concept of akratic action should also be noted here: Tappolet, 'Emotions and the Intelligibility of Akratic Action'.

[76] Russell, *Unpopular Essays*, 32.

[77] Russell, *History of Western Philosophy*, xiv.

[78] Russell explicitly identifies himself as an agnostic, in that he regarded the question of God to lie beyond proof: see especially Russell, *Essays in Skepticism*, 83–4; idem, *Bertrand Russell Speaks His Mind*, 20. However, Russell was prepared to allow that he was an atheist in the popular sense of that term, thus distinguishing his epistemological agnosticism from his pragmatic atheism.

other, are worrying signs of what is at best a disinclination, and at worst an unprincipled refusal, to take seriously the deeply troubling human predicament, so sensitively explored in Alexander Pope's *Essay on Man* (1732–4). For Pope, human beings are 'born but to die, and reas'ning but to err', inhabiting a middle realm between scepticism and certainty.[79] While the rationalities of religious fundamentalisms are somewhat more complex than their critics generally concede,[80] there is ample evidence of their quest for certainty in their convictions,[81] and their tendency to demonize those who do not share them.

These disturbing tendencies of religious fundamentalism are surprisingly mirrored, rather that countered, by some of their secularist counterparts. The populist 'New Atheism', which achieved media attention in 2006–7, secured its cultural traction largely by avoiding any serious engagement with the question of human rational capacities, preferring to offer grand unevidenced assertions of the essential rationality of scientific atheism and the fundamental irrationality of religious belief.[82] While it was unclear quite how the 'god' rejected by the New Atheism related to those of religious communities,[83] its arguments against religious belief often focussed on its perceived irrationality. For Christopher Hitchens, perhaps the most eloquent representative of this movement, the remedy for this resurgence of irrationality lay in rediscovering and returning to the elegant intellectual simplicities of the Golden 'Age of Reason', which he believed had been neglected, if not suppressed, in contemporary debates.[84]

The growing influence and aggressiveness of both religious fundamentalisms and their secular counterparts—such as the 'New Atheism'—might seem to confirm at least Russell's observation of

---

[79] Pope, *Essay on Man*, 38. For the intellectual context of Pope's ideas, see Cope, *Criteria of Certainty*, especially 140–67.

[80] For example, see the perceptive analysis in Euben, *Fundamentalism and the Limits of Modern Rationalism*, 20–48.

[81] For a critical assessment, see Vorster, 'Perspectives on the Core Characteristics of Religious Fundamentalism Today'; Hood, Hill, and Williamson, *The Psychology of Religious Fundamentalism*.

[82] Pigliucci, 'New Atheism and the Scientistic Turn in the Atheism Movement'. For the link between science and secularism, see Coleman, Hood, and Shook, 'An Introduction to Atheism, Secularity, and Science'.

[83] Bradley, Exline, and Uzdavines, 'The God of Nonbelievers'.

[84] Hitchens, *God Is Not Great*, 277–83.

the natural human longing for certainty, if not his judgement that it represents an intellectual vice. Yet it is clear that this potentially sterile debate might serve to open up more intellectually productive possibilities, in which the central and controlling notion of 'rationality' is subjected to critical analysis. It is clear that there is a need, catalyzed but not created by the debate over the 'New Atheism', to provide a scholarly exploration of how the concept of 'rationality' is understood and enacted in both science and religion, and how this might impact on their relationship as two of the most significant forces in contemporary culture. Might there be 'common resources of rationality'[85] which could serve as the basis for a dialogue, or even for mutual refinement of methods and enrichment of outcomes? Might there be some multi-perspectival approach to interdisciplinary dialogue, in which rationality is to be situated within the 'intersecting connections and transitions between disciplines'?[86]

Such discussions need to be located within the 'turn to practice' in contemporary cultural theory, which recognizes a reciprocal and socially constructed relationship between belief and practice.[87] This approach, which has secured significant traction within many disciplines, involves moving beyond the traditional conception of reason as an innate mental faculty and reconceptualizing it in terms of practice—for example, by exploring how different epistemic communities develop their own distinct rational procedures and forms of argumentation.[88] The influence of this development is now evident across disciplines—for example, in the attentiveness of recent writers in the philosophy of technology towards the actual practices of engineering, or contemporary approaches to the philosophy of religion which emphasize that religious practices are irreducibly cognitive, and hence merit philosophical attention.[89]

---

[85] I borrow this helpful phrase from van Huyssteen, *The Shaping of Rationality*, 111–77.
[86] Van Huyssteen, *Alone in the World?*, 20.
[87] See especially Schatzki, Knorr-Cetina, and von Savigny, eds, *The Practice Turn in Contemporary Theory*.
[88] This trend is often traced back to Toulmin, *Human Understanding*.
[89] See Franssen, Vermaas, Kroes, and Meijers, eds, *Philosophy of Technology after the Empirical Turn*; Schilbrack, *Philosophy and the Study of Religions*, 31–49.

This growing interest in practice is of particular importance in the natural sciences. Since the 1980s, there has been growing anxiety within the philosophy of science concerning the then-dominant narratives of scientific progress, which assumed that the fundamental methodological norms for the sciences—such as explanation and justification—were in all important respects constant and invariant across scientific disciplines in the present, and in their historical development.[90] In distancing themselves from such overstatements, philosophers of science have instead proposed domain-specific approaches to scientific rationality, in which scientific methods are not predetermined, fixed, or universal, but rather emerge through scientific practice in forms adapted to their specific domains or objects of investigation.

This approach to scientific rationalities has not yet filtered through into the self-understanding of many scientific popularizers, who tend to prefer the older 'central narrative'. However, this approach fits well with (although is not dependent upon) the growing tendency within the philosophy of science to see the diverse natural sciences in terms of practice—as the collective activity of scientists working in socially coordinated communities, thus functioning as active participants in a social enterprise.[91] A recognition of the interconnectedness of the cognitive, epistemic, and social dimensions of science does not, however, render the notion of 'science' incoherent; it is perfectly possible to develop models—such as that of 'family resemblances'—which allow individual scientific traditions to be located within a broader coherent framework of understanding.[92]

ONE REASON; MULTIPLE RATIONALITIES

We may indeed speak generically of human reason in the singular; it has, however, become clear that we must now also speak of multiple

---

[90] Examples of such works from the 1980s include Galison, 'History, Philosophy, and the Central Metaphor'; Miller, *Fact and Method*; Laudan, *Science and Values*; Shapere, *Reason and the Search for Knowledge*.

[91] For an important early intervention of importance to this development, see Skinner, 'A Case History in the Scientific Method'.

[92] Rocha and Gurgel, 'Descriptive Understandings of the Nature of Science'.

rationalities.[93] Empirical research projects, such as the 'Beyond Rationality' programme established by the Centre for the Philosophy of Natural and Social Science at the London School of Economics, have attempted to map domain-relative practices and discourses relating to rationality, highlighting the diversity of the notion.[94] This mapping of the territories of reason expresses an increased awareness of the existence of 'multiple situated rationalities',[95] rather than the older belief in a context-free concept of rationality (or of its presumed antithesis, irrationality).

Each academic discipline develops its own specific implementation of rationality, adapted to the objects of its enquiry, and often demanding the application of wisdom, craft, and judgement rather than the mechanical application of procedural formulae.[96] This can be framed in terms of disciplinary 'fields', which are distinguished by a distinct focal problem, a domain of facts related to the problem, explanatory goals, methods, and an associated vocabulary.[97]

Pierre Duhem's notion of *le bon sens* articulates these intellectual intuitions, which arise from accumulated experience within a given epistemic community.[98] Even within a single discipline, the norms of rational discourse and practice have varied across history and culture, and continue to diverge across disciplines—including the natural sciences. Historical scholarship indicates that the boundary between reliable and established knowledge on the one hand, and the realm of conjecture and speculation on the other, is constantly shifting, in line with the changing dynamic of research and interpretation.[99] The 'received wisdom' of a research tradition is subject to constant review and reassessment, with the result that 'the price of scientific progress is the obsolescence of scientific knowledge'.[100]

---

[93] See especially Apel and Kettner, eds, *Die eine Vernunft und die vielen Rationalitäten*.

[94] Harré and Jensen, eds, *Beyond Rationality: Contemporary Issues*.

[95] Avgerou, *Information Systems and Global Diversity*, 72–97.

[96] For a detailed study of this point in relation to particle physics, see Franklin, *Shifting Standards*.

[97] Darden and Maull, 'Interfield Theories'.

[98] Stump, 'Pierre Duhem's Virtue Epistemology'.

[99] Daston, 'Scientific Error and the Ethos of Belief'. For an earlier and influential statement of this point, see Polanyi, *Personal Knowledge*, 315–42.

[100] Daston, 'Scientific Error and the Ethos of Belief', 1.

By the middle of the 1980s, there was a growing appreciation of the resistance of the notion of rationality to conformity with precise standards or procedures of reasoning, partly because of the sheer variety of modes of reasoning observed within human history and culture, and partly because of the growing appreciation of the significance of the social construction of rationality, which called into question the notion of some 'natural' human rationality which was invariant across culture and history.[101] The concept of rationality was increasingly realized to be so 'vague' and 'disparately employed' across intellectual disciplines—such as philosophy or the social sciences—that it had limited utility.[102] It is necessary to speak at least of a 'Discours préliminaire', which is embodied in different forms in various intellectual disciplines and historical contexts, leaving open at this stage the question of whether and how these various rationalities are related to one another.

The psychologist Jerome Kagan notes that there are multiple issues involved in any attempt to clarify notions of rationality across the 'Three Cultures'—by which he means the natural sciences, the social sciences, and the humanities—including:[103]

1. The primary questions asked, including the degree to which prediction, explanation, or description of a phenomenon is the major product of enquiry;
2. The sources of evidence on which inferences are based and the degree of control over the conditions in which the evidence is gathered;
3. The vocabulary used to present observations, concepts, and conclusions, including the balance between continuous properties and categories and the degree to which a functional relation was presumed to generalize across settings or was restricted to the context of observation;
4. The degree to which social conditions, produced by historical events, influence the questions asked.

---

[101] Scholte, 'On the Ethnocentricity of Scientistic Logic'; Bouwmeester, *The Social Construction of Rationality*.
[102] Goldman, *Epistemology and Cognition*, 27. Cf. Code, 'On the Poverty of Scientism, or: The Ineluctible Roughness of Rationality'.
[103] Kagan, *The Three Cultures*, 1–50.

For Jerome, these three different cultures share a fundamental characteristic: their claims are considered to be valid and coherent only *within* their own scientific community, and are not seen as compelling or binding from the standpoint of the other epistemic communities. This, Jerome rightly notes, constitutes a formidable barrier to meaningful dialogue. Perhaps, it might be added, it makes it all the more difficult for a member of any one of these communities to expect to learn something useful from the other two.[104]

Cognitive scientists distinguish between rationality and intelligence, noting that the developmental trajectories of the cognitive skills that underlie intelligence and those that underlie rational thinking are conceptually and empirically separable, and must thus be studied in their own right.[105] Two distinct kinds of rationality are distinguished within cognitive science: epistemic and instrumental. Epistemic rationality is about forming true beliefs, about ensuring that human mental maps accurately and adequately reflect the territory of the world. Instrumental rationality is the optimization of the individual's goal fulfilment, and concerns making decisions that are adapted to bring about intended or desired outcomes.[106]

One model of rational judgement assumes that a person chooses options with the greatest expected utility. Such rational choice models, which are particularly influential in economic theory, are difficult to apply in other contexts, and tend to be dependent on certain ad hoc auxiliary assumptions about how this approach is to be applied to these different situations.[107]

In practice, however, human beings seem to violate their own criteria of rationality on a regular basis. A rich range of research in the heuristics and biases tradition has shown that people violate many of the proposed norms of rationality—for example, in displaying confirmation bias, in testing hypotheses inefficiently, in displaying preference inconsistencies, in failing to properly calibrate

---

[104] Kagan, *The Three Cultures*, 245–75.
[105] Stanovich, 'On the Distinction between Rationality and Intelligence'.
[106] Toplak, West, and Stanovich, 'Assessing the Development of Rationality'; Stanovich, *Rationality and the Reflective Mind*, 95–174.
[107] The uncritical use of 'Game Theory' in this respect is an excellent example of this problem: see Gueth and Kliemt, 'Perfect or Bounded Rationality?'; Zouboulakis, *The Varieties of Economic Rationality*.

degrees of belief, in over-projecting their own opinions on to others, and in allowing prior knowledge to become entangled with deductive reasoning.[108] The significance of emotional influences on supposedly rational decisions has been the subject of many recent empirical studies,[109] which call into question less reflective accounts of human rationality, such as those which achieved cultural dominance during the 'Age of Reason'. 'The rule that human beings seem to follow is to engage [rational thought] only when all else fails—and usually not even then.'[110] Nevertheless, this still leaves the door open for improvement of our rational judgements by identifying the biases and constraints that may lead us to make bad decisions, and amend our styles of reasoning accordingly, thus hence increasing our capacity to reach 'rational' conclusions.

One of the most influential accounts of the emergence of a distinctive western rationality is due to the sociologist Max Weber, who argued that this has involved the depersonalization or commodification of social relationships, the refinement of techniques of calculation, the extension of technically rational control over both natural and social processes, and the enhancement of the social importance of specialized knowledge. Others, however, have argued that a number of different approaches to rationality can be discerned within the 'Age of Reason'. Although some writers consider modernity to have been displaced by a more pluralist postmodernity, with a wide range of rationalities, many scholars now recognize the multiple variations of modernity, its ongoing and innovative pluralization, and its lack of a normative notion of rationality.[111] A seemingly simple 'modernity' gave way to a multiplicity of 'modernities' through a process of appropriation and re-embedding of the ideas and practices of modernity in local situations, generating a multiplicity of local modern rationalities.[112] Despite some difficulties and divisions, the 'multiple

---

[108] Stanovich and West, 'On the Relative Independence of Thinking Biases and Cognitive Ability'.

[109] See Haidt, 'The Emotional Dog and Its Rational Tail'; Pham, 'Emotion and Rationality: A Critical Review and Interpretation of Empirical Evidence'.

[110] Hull, *Science and Selection*, 37.

[111] See especially Arnason, 'The Multiplication of Modernity'; Eisenstadt, 'Multiple Modernities'; Browne, 'Postmodernism, Ideology and Rationality'.

[112] Arce and Long, eds, *Anthropology, Development and Modernities*, 1, 160.

modernities' thesis provides a promising and potentially productive theoretical approach to frame and interpretively explain the vast body of empirical knowledge that has accumulated in recent decades pointing to the inadequacy of the 'inevitable and homogenizing' version of modernization theory, not least in relation to the future of religion in a diverse 'modern' world.[113]

Such a recognition of 'multiple modernities' permits an appreciation of the creative role of cultural contingency, resulting in different cultural inflections and embodiments of rationality, thus complexifying the older notion of a single Enlightenment rationality shaped by the notion of the rational mastery of the natural world—a theme of critical importance to both the natural sciences and Christian theology, and to those who define 'postmodernity' in terms of the perceived deficiencies of their understanding of 'modernity'. The notion of multiple rationalities can be mapped on to the rather different territories of both multiple modernities and postmodernity, and should not be interpreted simply in terms of the latter.

This point should be borne in mind when evaluating the 'postfoundationalist' proposal for theological rationality developed by Wentzel van Huyssteen, who locates his approach between modernity and postmodernity, aiming to avoid the difficulties arising from both.[114] Not only does such an approach overlook the importance of distinct disciplinary rationalities, which are not easily mapped on to the proposed territories of modernity and postmodernity; it does not give due weight to the *complexity* of modernity, and its capacity to develop in divergent (though arguably correlated) manners, adapted to local cultural norms and situations.[115] We need to speak of *modernities*, rather than a single *modernity*. As the very different implementations of modernity in Japan, China, and the Arab world indicate, 'transitions to what we might recognize as modernity, taking place in different civilizations, will produce different results, reflecting

---

[113] Smith and Vaidyanathan, 'Multiple Modernities and Religion'.

[114] van Huyssteen, *The Shaping of Rationality*. For comment, see LeRon Shults, ed., *The Evolution of Rationality*; Reeves, 'Problems for Postfoundationalists'.

[115] See the discussion in Fourie, 'A Future for the Theory of Multiple Modernities'; Schmidt, 'Multiple Modernities or Varieties of Modernity?'; Lee, 'In Search of Second Modernity'.

the civilizations' divergent starting points'.[116] The complex territory of human rationality cannot be accommodated in terms of a simple dichotomy between modernism and postmodernism—or any other binary axis, such as foundationalism and postfoundationalism—leading to the construction of some third category which retains the strengths of each component while minimizing their weaknesses.

The German philosopher Wolfgang Welsch has argued that, given the specificity of rationalities to each domain of enquiry, there is a need to reassert the category of 'transversal reason' as a means of transcending and mediating between these various construals and implementations of rationality.[117] However, even allowing for the merits of Welsch's approach, we are still left with the problem of 'local' accounts of rationality, in which it proves problematic to determine who (and what) is rational in the absence of an informing context.[118] As many have noted, for example, the laws of nature may seem universal and univocal, but often have different meanings for different practitioners or local situations.[119] Even the natural sciences—widely regarded as the most universal of disciplines—developed, and continue to develop, what might be called local 'styles', reflecting how nations or communities believed the sciences ought to be implemented.[120]

Yet a pluralism of rationalities arises for reasons other than the specificity of academic disciplines. Perhaps the most important of these is the historic or cultural location of individuals or communities, which plays a major role in determining what they accept as reasonable, and in judging those which they find to be irrational. The reception of ideas—a process which involves both the interpretation of their meaning and an assessment of their local utility—is significantly shaped by the cultural context of those who receive them.[121]

---

[116] Taylor, 'Modernity and Difference', 367.
[117] Welsch, *Vernunft*, 366-7. Welsch here draws on the pioneering 1964 article of Guattari, 'La transversalité'. The idea of transversal rationality is also found in the writings of Wentzel van Huyssteen, who draws on the somewhat less rigorous account of the idea found in the writings of Calvin Schrag.
[118] Huggett, 'Local Philosophies of Science'.
[119] Galison, 'Material Culture, Theoretical Culture, and Delocalization', 677-81.
[120] See Henry, 'National Styles in Science'.
[121] See Livingstone, 'Science, Text, and Space'; Withers, *Placing the Enlightenment*, 136-48.

Recognizing the 'constitutive significance of place' in the production of meaning does not entail a descent into irrationalism or radical scepticism, but rather calls for a warranted attentiveness to the complex historical and cultural geography of human reason.[122] Human rationality is rooted in, and hence shaped by, the realities of human biological and social existence. These themes were anticipated by early pragmatists, particularly Charles Peirce and John Dewey, who explored how personal convictions and local contexts interacted with more universal features of human experience. This conversation clearly needs to be continued and extended, particularly in the light of a new interest in 'post-foundationalism' in the fields of science and religion, which often seems to be inattentive to earlier discussions which retain a potential for further development.

RATIONALITY, IDEOLOGY, AND POWER

To its critics, one of the most significant achievements of the Enlightenment has been to create a social imaginary of rationality which makes what is actually culturally *contingent* appear to be intellectually *necessary*. This naturally leads into a discussion of one of the most important and difficult aspects of the concept of rationality—the manner in which human perceptions of what is 'reasonable' can be manipulated by power groups.

The term 'ideology'—minted in conscious opposition to 'theology'—was developed at the time of the French Revolution, which articulated its Revolutionary rationalism in terms of a science of ideas, founded on an objective system of classification which permitted clear and distinct lines to be drawn between truth and falsehood. For Antoine Destutt de Tracy, who is widely regarded as crystallizing the notion of ideology in the immediate post-Revolutionary period, there was an essential coincidence of ideology and rationality,[123] with the former

---

[122] Rupke, 'A Geography of Enlightenment', 336.
[123] Steger, *The Rise of the Global Imaginary*, 19–43; Freeden, *Ideology*, 4. Cf. Gusdorf, *La conscience révolutionnaire*. For de Tracy, ideology was a science that served as the basis for the critique of false or irrational ideas, and was not itself understood in terms of false ideas.

being understood as a systematic attempt to establish an empirically based science of ideas, capable of sustaining the cultural hegemony and intellectual coherence of the Revolutionary government.[124] Ideology was, in effect, a science of ideas which was created and sustained by those in political power, a historical contingency of this Revolutionary age.

The collapse of Revolutionary rationalism following Napoleon Bonaparte's accession to power ended the social privilege of this allegedly universal science of ideas, and led to ideology's being seen as a remnant of a discredited political past.[125] Yet its intellectual credibility was under threat for other reasons by the 1830s, in that there was a growing realization that there were a series of possible ideologies, rather than one natural and hence authoritative ideology.[126] There is no doubt that Karl Marx's *Ideologiekritik* was of fundamental importance in highlighting the social conditioning of ideologies and their cognate concepts of rationality. For Marx, an ideology was fundamentally a deliberate intellectual falsification or distortion of material reality which allowed the ruling classes to exploit workers economically, and oppress them politically. An ideology was a socially constructed set of ideas, a social imaginary designed to conceal class oppression and exploitation by demanding that the world be seen through what was seemingly a rational and self-evident interpretative framework, yet was in reality a constructed social imaginary designed to reinforce the hegemony of a specific social class.[127] In effect, an ideology represented an 'attempt to represent the universal from the particular point of view of the dominant class' in such a way that this artificial social construction seemed eminently reasonable

---

[124] For the role of lawmakers in embedding ideology in social structures, see especially Bauman, *Legislators and Interpreters*, 96–109.
[125] Steger, *The Rise of the Global Imaginary*, 1–2.
[126] For the issues, see Millon-Delsol, 'La dénaturation de la vérité ou le fondement des idéologies'.
[127] Márkus, 'Concepts of Ideology in Marx'. This is a leading concern in Castoriadis, *L'institution imaginaire de la société*. Cf. Taylor, *Modern Social Imaginaries*, 23–30. The theme is also important in Gramsci's account of the relation of ideology and culture: see especially Filippini, *Using Gramsci*, 4–23.

and natural.[128] There are obvious parallels between ideologies and religion, in that both offer 'big pictures' of reality which deal with core human concerns such as meaning, values, and purpose.[129] In the case of Christianity, a way of seeing the world is developed which is rooted in historical particularities—such as the history of Jesus of Nazareth—yet is interpreted as possessing universal significance.

The category of ideology thus allows us to explore the complex relationship between rationality and power—or, more accurately, the relationship between societies and the forms of rationality which they develop as a means of expressing and defending their core assumptions and social structures. They are culturally contingent interpretative frameworks which present a view from *somewhere* as if it were a view from *nowhere* (and hence disinterested and objective) or a view from *everywhere* (and hence universal). In one sense, an ideology is essential if a movement or society is to have some degree of intellectual or cultural coherence; in this somewhat neutral sense, ideology is inevitable and relatively unproblematic. The main concern of critics of ideology is not the phenomenon as such, which is arguably inescapable, given the human need to develop means of understanding and structuring the social and political worlds they inhabit.[130] Rather, the concern is with those ideologies which are deliberately constructed—one might say *fabricated*—to advance the agendas and influence of sectarian interests or specific social groupings. It is not so much that reason is abused in the service of vested interests; Marx's point is rather that such notions of rationality are often shaped by these vested interests in the first place.

These cautionary observations are not intended in any way to detract from the importance of the theme of rationality in human life in general, or in the fields of the natural sciences or Christian theology in particular. They are rather intended to frame an

---

[128] Lefort, *The Political Forms of Modern Society*, 200. For Lefort, an ideology controls the way we see the world, in that it 'establishes the origins of facts, encloses them in a representation and governs the structure of the argument' (205).

[129] Note the emphasis on the 'meaning dimension' in Clayton, *Explanation from Physics to Theology*, 113–45.

[130] See the important essay of Jos, Kay, and Thorisdottir, 'On the Social and Psychological Bases of Ideology'.

informed discussion which avoids abstract intellectual idealizations of rationality by exploring the forms which it actually has taken across cultures and disciplines in an empirical and non-judgemental way. This procedure establishes a context within which a more detailed comparison can subsequently be made between the enactments of rationality encountered within the natural sciences and Christian theology, which might offer possibilities of interdisciplinary enrichment.

This book aims to explore the notions of rationality encountered in the natural sciences and Christian theology, and reflect on how these might help us frame the intellectual enterprise of interdisciplinary dialogue, particularly the broader and more diffuse field usually referred to as 'science and religion'. Yet it is impossible to consider the relationship between the natural sciences and Christian theology without acknowledging that, like conceptions of rationality themselves, understandings of what the terms 'science' and 'religion' denote are embedded in cultural contexts, and inevitably reflect the agendas of those sociological addresses. Since religion is not an empirical notion, definitions of 'religion' inevitably reflect covert assumptions about what religion *ought* to be, and often unacknowledged judgements as to whether it is to be seen as benign or pathological.[131] This work cannot avoid such entanglements; it is, however, hoped that the approach it adopts will both acknowledge their significance, and minimize their intrusiveness.

---

[131] Harrison, 'The Pragmatics of Defining Religion in a Multi-Cultural World'; D'Costa, 'Whose Objectivity? Which Neutrality?'

2

Mapping Human Reason

Rationalities across Disciplinary Boundaries

The present study is conceived as an exploration of concepts of rationality across disciplinary boundaries,[1] focussing particularly on the porous intellectual borders between the natural sciences and Christian theology. Three main considerations lie behind the choosing of this boundary for particular consideration:

1. In the last forty years, the field of science and religion has emerged as a significant area of academic research and teaching, with leading universities establishing research centres in the field—such as the Ian Ramsey Centre at Oxford University, founded in 1986. Intensive research in the field has indicated the need for significant revision of scholarly understandings of the relationship of science and religion,[2] at the levels of both historical research and contemporary engagement and dialogue.

2. Science and religion remain highly significant elements of global culture, although with significant variations in how their mutual

---

[1] See, for example, the general discussions in Stenmark, *Rationality in Science, Religion, and Everyday Life*; Elio, ed., *Common Sense, Reasoning, and Rationality*; Bertolotti, *Patterns of Rationality*. On the specific case of science and religion, see Clayton, *Explanation from Physics to Theology*; Banner, *The Justification of Science and the Rationality of Religious Belief*; van Huyssteen, *The Shaping of Rationality*.

[2] Two works of fundamental importance in catalysing this revision of existing understandings are Brooke, *Science and Religion*; Harrison, *The Territories of Science and Religion*.

relationship is understood in local contexts,[3] reflecting differing degrees of religiosity, varying levels of scientific infrastructure, and unique relationships between religious and state institutions.

3. Christian theology has long recognized the importance of dialogue and debate with other systems of thought as a means of self-correction and self-improvement on the one hand, and as affording an enhanced capacity to communicate its ideas beyond its own boundaries.[4] The academic discussion of the nature and norms of rationality is clearly of theological significance, not least because certain schools of theology seemed to be wedded to outdated notions of the concept, and their implementation in practice.

Serious discussion of the place of rationality in science and religion has been impaired by the perpetuation of outdated stereotypes based on spurious binaries on the one hand, and oversimplifications and distortions on the other—as evidenced in the following bizarre assertion.[5]

Being scientifically rational requires seeking evidence, especially evidence refuting a specific belief. It requires taking this evidence seriously and changing our beliefs if the evidence demands it. Being religious requires us to ignore evidence, especially refuting evidence.

Such unevidenced overstatements may have a place in popular debates, in which complexification and engagement with alternative perspectives are seen as needless impairments obscuring a self-evident truth, whose luminosity obviates any need for assessment or critical reflection. Yet even a superficial familiarity with religious works on the rationality of belief is sufficient to discredit such superficial oracular pronouncements,[6] which ought to have no place in serious discussion of such issues.

It would be impossible for a single monograph to engage the topic of rationality across the vast territories of science and religion in a

---

[3] See the recent detailed analysis covering eight geographical regions in Ecklund et al., 'Religion among Scientists in International Context'.

[4] For reflections on this notion, see Detjen, *Geltungsbegründung traditionsabhängiger Weltdeutungen*, 201–7.

[5] Dietrich, *Excellent Beauty*, 106.

[6] For example, see Davies, *Thinking about God*, 235–305; McGrath, 'The Rationality of Faith'.

manageable and meaningful manner; the present study therefore focusses on the more limited question of rationality as understood and practised within Christian theology and a representative sample of the natural sciences, especially the physical and biological sciences. It is a field in which there is already considerable interest—evident, for example, in the growing scholarly focus on the theological foundations of medieval and early modern science and mathematics on the one hand,[7] and on the potential value of the natural sciences in discussions about theological method.[8]

As we noted earlier (10–11), difficulties encountered in defining 'religion' cannot be entirely overcome,[9] not least because religion is a socially constructed, not an empirical, category. The problematic distinction between 'magic' and 'religion',[10] which is particularly significant in assessing the rationality of religion and cognate phenomena in the Middle Ages and Renaissance,[11] should also be noted. The lingering cultural stereotype of the essential 'irrationality' of religion also casts a lengthening, if ultimately unscholarly, shadow over the proposed topic of this work. Individual theologians who have criticized the ultimate authority of reason—such as Martin Luther— are sometimes misrepresented, especially in popular polemical literature, as abandoning any attempt to engage in rational discourse.[12]

---

[7] There is a large literature. Representative studies include Lawrence and McCartney, eds. *Mathematicians and Their Gods*; Gaukroger, 'The Early Modern Idea of Scientific Doctrine and Its Early Christian Origins'; Hon, 'Kepler's Revolutionary Astronomy'.

[8] There is a large literature. Representative studies include Torrance, *Theological Science*; Polkinghorne, *Scientists as Theologians*; McGrath, *Enriching our Vision of Reality*.

[9] Fitzgerald, 'A Critique of Religion as a Cross-Cultural Category'. See also the more general analysis of the changing cultural understanding of religion set out in Harrison, *The Territories of Science and Religion*, 21–54.

[10] Keith Thomas, for example, suggested that religion primarily offered an explanation of human existence while magic was more concerned with specific temporary problems: Thomas, *Religion and the Decline of Magic*. See further the important discussions in Sanchez, *La rationalité des croyances magiques*, and more generally Josephson-Storm, *The Myth of Disenchantment*, 4–16.

[11] Kieckhefer, 'The Specific Rationality of Medieval Magic'; Coudert, *Religion, Magic, and Science in Early Modern Europe and America*, 25–44.

[12] See, for example, Dawkins, *The God Delusion*, 190, based on some unreliable web sources. For a good account of Luther's somewhat more complex views on this matter, see Kern, *Dialektik der Vernunft bei Martin Luther*, 335–403.

Tertullian is often ridiculed for his aphorism *credo quia absurdum*—despite the obvious difficulty that this represents a reductive paraphrase of his rather more complex and nuanced argument.[13] Kierkegaard's selective appropriation of Tertullian in developing his own views on the irrationality of faith, though not without interest,[14] fails to do justice to the latter's emphasis on the rationality of faith, carefully framed to avoid cultural accomodationism.[15]

The present work cannot entirely avoid entanglements with the historical past, nor with continuing debates about the distinctive characteristics of both science and religion. The approach adopted, however, is intended to minimize these difficulties, by focussing primarily on the question of how the biological and physical sciences on the one hand, and Christian theology on the other, have construed and implemented the notion of rationality since the middle of the nineteenth century. The approach adopted is to offer a comparative account of how these two traditions of enquiry deal with a number of specific issues, traditionally seen as reflective of their commitment to the rationality of their undertakings—such as the role of evidence-based reasoning, the criteria of theory choice, and the manner in which anomalies are accommodated within theoretical structures.

ON THE CORRELATION OF RATIONALITIES

We have already noted the growing realization that rationality is domain-specific, in that each academic discipline develops a set of intellectual virtues, procedures, and criteria which are deemed to be appropriate to its distinct tasks and procedures. While human rational investigation of our world may be 'universal in intent', the rationality of particular beliefs and actions is generally 'person- and

---

[13] Sider, 'Credo Quia Absurdum?'
[14] Bühler, 'Tertullian: The Teacher of the *Credo Quia Absurdum*'.
[15] See, for example, Tertullian, *de poenitentia* I, 2. 'Quippe res dei ratio quia deus omnium conditor nihil non ratione providit disposuit ordinavit, nihil enim non ratione tractari intellegique voluit'. Tertullian tries to avoid equating *ratio* with culturally dominant notions of 'common sense'.

situation-relative'.[16] This is not a new development. For example, the Spanish philosopher Vicente Fernández Valcárcel (1723-98) criticized the universalizing tendencies of some Enlightenment writers, holding that 'each subject has its method (*cada materia tiene su método*)', enfolding its own distinct epistemological 'jurisdiction (*jurisdicción*)'.[17]

So how do these multiple rationalities relate to one another? Is there some 'meta-rationality' which they instantiate, each in its own distinct manner? Or are they to be seen as essentially independent, perhaps bearing some family resemblance allowing the door to be kept open for at least the *possibility* of shared norms or methods across disciplines? And what form might such a meta-rationality take? Are we speaking of a standpoint of logical necessity, based, like Euclid's geometry, on axiomatic deduction? Or of a plausible cohesiveness, holding together a domain of particulars after the manner of James Joyce's *Ulysses*?

The case of Isaac Newton—a leading figure in the Scientific Revolution of the late seventeenth century—is illuminating. Newton actively engaged the fields of natural philosophy and Christian theology, regarding both as constitutive elements of his intellectual and spiritual identity.[18] Yet there is no evidence of a single rational methodology underlying both these enterprises, which appear to have been conducted on the basis of quite different working assumptions. Where some older studies of Newton ambitiously proclaimed that both were the outcome of a 'single mind', the most recent study of Newton's religious thought has indicated that the documentary evidence provides 'no support for the notion that there is some simple conceptual or methodological coherence to his work'.[19] Newton clearly worked with distinct, even divergent rationalities, holding them together (for we cannot speak meaningfully of 'integration' here) in a manner that was peculiar to him as an individual, allowing

---

[16] van Huyssteen, *Shaping of Rationality*, 155.
[17] Valcárcel, *Desengaños filosóficos*, vol. 1, 408.
[18] Snobelen, 'To Discourse of God'; Illiffe, 'Newton, God, and the Mathematics of the Two Books'.
[19] Illiffe, *Priest of Nature*, 14.

the reconciliation of such approaches in an idiosyncratic manner that is probably inaccessible to us.

A further issue that might be raised here is the willingness of an epistemic community to justify itself to a wider public. Advocates of a public Christian theology, for example, argue that the theologian must always critically reflect on Christian practice in dialogue with those inside and outside the community. 'Any appeal to hidden or private sources of authority or justification is inappropriate for a genuinely public theology. The structure and logic of theological argument must be available for examination by any reasonable inquirer.'[20] This concern for rational and evidential accountability across traditions may be complexified by the recognition of multiple operating rationalities; it is not, however, invalidated by it.

Discussion of this already difficult interdisciplinary question of the correlation of rationalities is made more intractable by the principled refusal of some in both the scientific and the theological camps to sanction any meaningful conversations or discussions between their respective communities.[21] Some draw on the deeply flawed—but still widely cited—taxonomy of possibilities advocated by Ian Barbour,[22] two elements of which advocate either a perennial 'warfare' between the natural sciences and Christian theology or their total intellectual independence, to suggest that interdisciplinary conversations will necessarily be sterile and unproductive. Both sides are prone to use the toxic language of contamination, concerned that their own distinct insights, methods, and values might be tainted by intellectual contact with alternative ways of thinking. There is a reluctance, perhaps even a failure, to acknowledge the possibility of intellectual enrichment within both intellectual communities, probably because this entails an implicit recognition of incompleteness and penultimacy of both the natural sciences and theology.[23]

Although such attempts to avoid engaging with the natural sciences are widely encountered within Christian theology—especially

---

[20] Thiemann, *Constructing a Public Theology*, 20. See further Chung, *Postcolonial Public Theology*, 21–125.
[21] O'Brien and Noy, 'Traditional, Modern, and Post-Secular Perspectives on Science and Religion in the United States'.
[22] Cantor and Kenny, 'Barbour's Fourfold Way'.
[23] For an example of this approach, see McGrath, *Enriching our Vision of Reality*.

in relation to psychology[24]—some sections of the scientific community are also resistant to serious engagement with other disciplines, or to engaging their own limitations. Many would now argue that the practices of the natural sciences are too easily reified into something that gives the impression of being inevitable and perennial, and in doing so becomes inattentive to its own history and contingency, in effect having become 'a sealed microworld marked by fixity, holism and cohesiveness, and isolated from everything and everyone else by a mathematically thin but poreless boundary'.[25] The pressure for a clear demarcation between 'science' and 'non-science' is often presented as arising from the need to demarcate a zone of credibility within which reliable knowledge is created; the social scientist, however, is more likely to see this as the creation, defence, and policing of culturally contingent and shifting boundaries, which ultimately reflect changing scientific practice.[26]

Any suggestion that both the natural sciences and Christian theology might enrich their grasp of our immense and complex universe, including ultimate questions of meaning, value, and purpose, through interaction and dialogue thus falls on many deaf ears within both the scientific and religious communities. Any suggestion of incompleteness on the part of science or religion is seen as entailing their inadequacy. This is, it need hardly be said, a misreading of the situation. The distinct approaches of both the natural sciences and Christian theology are such that they are *necessarily* incomplete; they can only be made to be complete by some form of metaphysical inflation which compromises each of their distinct identities and research methods.

SCIENTISM: THE NATURAL SCIENCES AS THE ULTIMATE RATIONAL AUTHORITY

This resistance to dialogue, ultimately grounded in a misplaced fear of intellectual contamination, is heightened by what some have called

---

[24] See the detailed analysis in Collicutt, 'Bringing the Academic Discipline of Psychology to Bear on the Study of the Bible'.
[25] Galison, 'Scientific Cultures', 128.
[26] Gieryn, *Cultural Boundaries of Science*, x–xii.

'scientism'—an inelegant contraction of 'scientific imperialism'—which privileges the natural sciences, holding that scientific enquiry enables the resolution of conflicts and dilemmas in contexts where traditional sources of wisdom and practical knowledge are seen to have failed.[27] The philosopher Ian Kidd has argued that three basic 'impulses' can be discerned as lying behind the rise of scientism:[28]

1. An *imperialist urge*—a compulsion to extend the concepts, methods, and practices of scientific enquiry into areas in which their competency is at best uncalibrated, and almost certainly problematic.

2. A *salvific urge*—an insistence that science, or what some people *take* to be science, can satisfy our ethical, spiritual, and existential concerns and needs.

3. An *absolutist urge*—a compulsion to assign to science the exclusive task of providing complete, absolute, and 'totalizing' accounts of life, the universe, and everything.

Scientism has thus gradually come to be understood as 'a totalizing attitude that regards science as the ultimate standard and arbiter of all interesting questions; or alternatively that seeks to expand the very definition and scope of science to encompass all aspects of human knowledge and understanding'.[29] These ideas are echoed in popularist defences of scientism, such as the somewhat brash statements of Peter Atkins: 'Reductionist science is omnicompetent. Science has never encountered a barrier that it has not surmounted.'[30]

Such forms of scientism are however, vulnerable to penetrating criticisms. It fails to account for the astonishing success of mathematics, which does not derive its ideas through scientific means, even if those ideas may ultimately prove to have scientific utility. More importantly, scientism finds itself trapped in a viciously circular argument from which no experiment can extricate it, in that it has

---

[27] Dupré, *Human Nature and the Limits of Science*, 74. See further Robinson and Williams, eds, *Scientism: The New Orthodoxy*; Stenmark, *Scientism*; McGrath, 'Gli ateismi di successo'.
[28] Kidd, 'Doing Science an Injustice'.
[29] Pigliucci, 'New Atheism and the Scientistic Turn in the Atheism Movement', 144.
[30] Atkins, 'The Limitless Power of Science', 129.

to assume its own authority in order to confirm it. Inflated forms of scientism, which treat science as the 'ultimate standard and arbiter of all interesting questions' are actually making second-order philosophical claims about science, which cannot be verified empirically; a refutation of this point must therefore rest on philosophical, not scientific, arguments. The price of getting out of this vicious circle is forfeiting such spurious claims to intellectual privilege.[31]

To break out of this circle requires 'getting outside' of science altogether and—discovering from that extra-scientific vantage point that science conveys an accurate picture of reality—and, if scientism is to be justified, that only science does so. But then the very existence of that extra-scientific vantage point would falsify the claim that science alone gives us a rational means of investigating objective reality.

Scientism is perceived to be arrogant by non-scientists. For the philosopher Mary Midgley, 'scientism's mistake does not lie in over-praising one form of [knowledge], but in cutting that form off from the rest of thought, in treating it as a victor who has put all the rest out of business'.[32] Yet the real problem is that the rich variety of human discourses and experience prove resistant to even the most persistent demands that they should be reduced to any single vocabulary,[33] whether this be scientific or something else. Midgley insists that most of the important questions in human life demand a number of different conceptual tool-boxes that need to be used together.[34] If a single perspective on reality is allowed to become normative, the outcome is inevitably a 'bizarrely restrictive view of meaning'.[35] Midgley's approach recognizes the need for 'multiple maps' of a complex reality. No single approach is adequate; different angles of approach and research methodologies are required in order for the human mind to secure a maximal grasp of the universe.

---

[31] Feser, *Scholastic Metaphysics*, 10–11.
[32] Midgley, *Are You an Illusion?*, 5. Susan Haack frames scientism in terms of attitudes of 'exaggerated deference' to, or an 'excessive readiness' to revere, science: Haack, *Defending Science*, 17–18.
[33] Putnam, *Naturalism, Realism, and Normativity*, 42–7. Cf. his critique of 'pan-scientism': *Naturalism, Realism, and Normativity*, 214.
[34] Rivera, *The Earth is Our Home*, 179.
[35] Midgley, *Wisdom, Information, and Wonder*, 199.

Mapping Human Reason

In this chapter, I propose to map two approaches to conceptualizing the rational relationship between science and religion in general (and science and Christian theology in particular). Both are essentially heuristic, offering an imaginative framework which helps us to see how their intellectual territories might overlap and interact, without providing a rigorous means of adjudicating territorial boundary disputes or determining levels of porosity. These can easily be used to facilitate a wider transdisciplinary dialogue.

MULTIPLE PERSPECTIVES ON A COMPLEX REALITY

Earlier, I cited the biologist Steven Rose, who holds that we live in a world that is an 'ontological unity', while recognizing that we must adopt 'an epistemological pluralism' in investigating it.[36] In defending and unfolding this view, Rose offers us a 'fable' of five biologists, representing different subdivisions of that discipline, who observe a frog jump into a pond. What explanations might be offered of this observation?

The physiologist explains that the frog's leg muscles were stimulated by impulses from its brain. The biochemist supplements this by pointing out that the frog jumps because of the properties of fibrous proteins, which enabled them to slide past each other, once stimulated by ATP. The developmental biologist locates the frog's capacity to jump in the first place in the ontogenetic process which gave rise to its nervous system and muscles. The animal behaviourist locates the explanation for the frog's jumping in its attempt to escape from a lurking predatory snake. The evolutionary biologist adds that the process of natural selection ensures that only those ancestors of frogs which could detect and evade snakes would be able to survive and breed. Rose's point is simple: all five explanations are part of a bigger picture. All of them are right; they are, however, different. 'The

---

[36] Rose, 'The Biology of the Future and the Future of Biology', 128–9.

most that we can insist is that explanations in the different discourses should not contradict each other.'[37]

Rose's fable illustrates the issues that need to be considered in moving from the recognition of multiple perspectives to the development of a unified theoretical account. Each of the five approaches can be treated as offering a specific a perspective on the frog's jump. The spatial metaphor does not require that these be treated as 'fictions', or even as instrumentalist accounts of the phenomenon. They are *perspectives*, reflecting their own distinct disciplinary methods and emphases.

The idea that there are multiple ways of viewing or approaching a complex reality can be traced back to Plato. Reality is too vast to be fully apprehended by any single individual; we can at best hope to grasp part of that greater whole, and allow others to supplement our limited apprehension. Knowledge is thus a communitarian or corporate undertaking, involving the aggregation and assimilation of multiple perceptions. C. S. Lewis is perhaps one of the best-known representatives of this view, arguing that literature represents an accumulation of insights, open to personal appropriation and synthesis. 'My own eyes are not enough for me, I will see through those of others ... In reading great literature, I become a thousand men and yet remain myself. Like the night sky in the Greek poem, I see with a myriad eyes, but it is still I who see.'[38] Literature, for Lewis, enables us 'to see with other eyes, to imagine with other imaginations, to feel with other hearts, as well as our own'.[39] The weaving together of such multiple perspectives and partial insights is left to the creative imagination of the individual knower.[40]

Nietzsche developed a related approach in his *Will to Power*, arguing that the human eye cannot take in the rich landscape with which it is confronted, and tends to focus on features in the immediate foreground of the field of vision. This complexity of the visual field is such that we are unable to accommodate it in terms of a single meaning (*Sinn*) lying behind the world; rather, there are countless

---

[37] Rose, 'The Biology of the Future and the Future of Biology', 128.
[38] Lewis, *An Experiment in Criticism*, 140–1.
[39] Lewis, *An Experiment in Criticism*, 137.
[40] See further McGrath, 'An Enhanced Vision of Rationality'.

meanings.⁴¹ Although the basis and implications of Nietzsche's perspectivism are the subject of considerable debate, it seems that the decision to accept one such meaning as normative is thus best seen as an act of intellectual self-determination rather than of discernment.⁴² Truth is not something that is 'there', waiting to be 'found or discovered', but is rather 'something that must be created', and that thus refers to a *process*, or a 'will to overcome that has in itself no end'. Nietzsche himself does not develop an intellectual project designed to integrate multiple perspectives, being more concerned with exploring the implications of the human tendency to identify what is seen to be relevant for pragmatic reasons. Yet the depth of Nietzsche's analysis seems to require him to work with—or at least presuppose—some form of ontological perspectivalism.⁴³

The philosopher of science Ronald J. Giere argues that some form of 'perspectival realism' is now essential in order to construct a middle way between two flawed conceptions of the natural sciences:⁴⁴ *classical science*, framed by the assumptions of the Enlightenment, which sees the natural sciences as an objective and rational pursuit of the laws of nature; and *social constructivism*, which holds that science is a social process, and as such, cannot be separated from human values and beliefs. Giere offers a more nuanced and behaviourally sophisticated approach that finds a middle ground between classical and social constructivist views of science, both acknowledging that science is indeed a social process, and insisting that scientific models nevertheless bear an objective relation to the world.

In his account of how perspectival models can lead to theory unification,⁴⁵ Alexander Rueger adopts an approach similar to Giere's 'perspectival realism',⁴⁶ noting that perspectival models '[are] not restricted to models incorporating a spatial perspective (which provides for the analogy with visual perspectives)', but can be extended

---

⁴¹ Nietzsche, *Sämtliche Werke*, vol. 9, 337: '[Die Welt] hat keinen Sinn hinter sich, sondern unzählige Sinne'.
⁴² Cf. Tanesini, 'Nietzsche's Theory of Truth'; Heelan, 'Nietzsche's Perspectivalism'.
⁴³ As argued by Welshon, 'Perspectivist Ontology and *De Re* Knowledge'.
⁴⁴ Giere, *Science without Laws*; idem, *Scientific Perspectivism*. For criticism of this general approach, see Votsis, 'Putting Realism in Perspective'.
⁴⁵ Rueger, 'Perspectival Models and Theory Unification'.
⁴⁶ Giere, *Scientific Perspectivism*.

to include the recognition of multiple *levels* of a system. In many cases, these different perspectives yield a complementarity which allows for the development of a coherent treatment of the phenomena; in some others—such as attempts to model atomic nuclei—the models appear inconsistent, and the development of a unified theory proves difficult, if not impossible.[47] Yet it is clear that this analogy has potential to illuminate one of the most important aspects of any form of transdisciplinary research—the capacity to bring together 'information, data, theories, and methodologies from multiple disciplinary viewpoints' in order to 'create something new that is irreducible to the disciplinary components that were initially brought to bear'.[48]

Defenders of certain forms of scientific perspectivalism can thus argue that the approach is able to describe not merely different regions of the same complex system but the same system at different levels.[49] Perspectives are to be seen as a visual metaphor, an imaginatively rich yet cognitively inexact manner of describing the many features of a complex system, without necessarily resolving the complexities of their relationships. The appeal to multiple perspectives is a strategy for saving the phenomena, providing a conceptual net that captures complexity and detail, yet without resolving the relationship of the various elements of the picture. No perspective offers a 'perfect model' of reality; rather, each perspective offers an account of reality which is not that of exact *isomorphism*, but rather that of *similarity*, and then always in limited respects and degrees.[50] The challenge is that of the colligation—perhaps even unification—of such perspectives without loss of their local explanatory power.[51]

The visual metaphor of 'perspective' thus offers more than the recognition of multiple ways of viewing and representing a complex reality; it also catalyses discussion about different *levels* of reality, by offering a means of visualizing the depth of a complex entity. The origins of linear perspective at the time of the Renaissance arose from an artistic desire to be able to convey depth in drawings, thus enabling

---

[47] See the detailed discussion in Morrison, 'One Phenomenon, Many Models'.
[48] Leavy, *Essentials of Transdisciplinary Research*, 31.
[49] See the important analysis in Rueger, 'Perspectival Models and Theory Unification', especially 590–2.
[50] For this point, see Teller, 'Twilight of the Perfect Model Model', 395–402.
[51] Rueger, 'Perspectival Models and Theory Unification'.

two-dimensional representation of a three-dimensional reality.[52] Yet there are concerns that need to be raised about such an approach, including the anxiety that the use of perspective introduces a homogeneity or orderedness which is alien to the direct experience of reality itself, thus imposing a predetermined structure on what is being observed.[53]

SCIENCE AND THEOLOGY: DISTINCT PERSPECTIVES ON REALITY

The notion that science and religion offer different, yet potentially complementary, perspectives on reality has been widely explored.[54] The Oxford theoretical chemist Charles A. Coulson used this image in a number of his writings, seeing it as a heuristic device which allowed the fundamental coherence of science and faith to be affirmed.[55] Coulson's personal interest in relating science and faith, which emerged during his time as an undergraduate at Cambridge University, led him to refuse to think in terms of a 'dichotomy of existence' between science and faith, as if our experience of the world could be pre-assigned to self-contained and mutually agonistic 'religious' or 'scientific' categories. He was not prepared to countenance the notion of 'some sort of hedge in the country of the mind' that separated these two domains.[56] Coulson was highly critical of any view that held that it was possible to allocate different intellectual locations for science and religion, or to regard these as domains which were under separate authority.

---

[52] For the suggestion that there are theological roots to this approach, see Edgerton, *The Mirror, the Window and the Telescope*, 36.

[53] See especially Panofsky, 'Die Perspektive als "Symbolische Form"'. Various recent artistic movements have arisen as protests against such a rigid representation of a complex and changing reality: see Hendrix, *Platonic Architectonics*, 175–202.

[54] See especially Watts, 'Science and Theology as Complementary Perspectives'. For related approaches, see Wolterstorff, 'Theology and Science'; Rueger, 'Perspectival Models and Theory Unification'.

[55] Coulson, *Christianity in an Age of Science*; idem, *Science and Christian Belief*; idem, *Science and the Idea of God*.

[56] Coulson, *Science and Christian Belief*, 19.

As an enthusiastic mountaineer, Coulson illustrated his own approach by inviting his readers to imagine Ben Nevis, Scotland's highest mountain. Seen from the south, the mountain presents itself as a 'huge grassy slope'; from the north, as 'rugged rock buttresses'. Those who know the mountain are familiar with these different perspectives. 'Different viewpoints yield different descriptions.' A full description of the same mountain requires these different perspectives to be brought together, and integrated into a single coherent picture.[57] The scientist might thus stand at the north side of the mountain, the poet at the south, and so on. Each reports on what they find using their own distinct language and imagery, adapted to what they see. Where one observer might see grassy slopes, another might see a rocky mountain. Yet both are representative and legitimate viewpoints of the same greater reality. For Coulson, this makes the need for an overall, cumulative, and integrated picture of reality essential. 'Different views of the same reality will appear different, yet both be valid.'[58]

Coulson refuses to allow that there are demarcated 'scientific' and 'religious' worlds, which are each experienced in different manners. It is one and the same world that is experienced—and that experience is complex, requiring and mandating both scientific and religious approaches. 'The two worlds are one, though seen and described in appropriate terms; and it is only the man who cannot, or will not, look at it from more than one viewpoint who claims an exclusive authority for his own description.'

There are clearly weaknesses with this general approach, most notably the need to negotiate intellectual boundaries and methodological privileges. Are all standpoints or angles of approach to be regarded as being of equal value and utility? Or might one serve in effect as a privileged or normative standpoint, offering a conceptual framework on the basis of which others might be evaluated, positioned, and coordinated?

A second concern relates to the inability of such an approach to do justice to the complex texturing of reality. How can such perspectives account for the *granularity* of reality? Much attention has been paid

---

[57] Coulson, *Christianity in an Age of Science*, 20.
[58] Coulson, *Christianity in an Age of Science*, 21.

to this problem in recent years, especially the notion of 'granular perspectives', and the related question of the processes of cognitive partitioning of the world which underlies the quest for multiple veridical perspectives.[59]

As it stands, then, Coulson's perspectival approach reflects a very 'flat' view of the world, which seems inattentive and insensitive to the possibility of multiple levels or 'strata' of reality. For example, religion is a complex social phenomenon, which possesses and is characterized by multiple dimensions or levels.[60] So how can we do justice to the multiple layers of religion—such as its symbols, narratives, practices, and virtues?

We have already seen how the notion of linear perspective opens up one possibility of representing depth while working with two-dimensional representative forms. Galileo's knowledge of this approach allowed him to correctly interpret his telescopic observations of the moon as disclosing the presence of such features as mountains—despite the fact that his observations were limited to two dimensions.[61] Strictly speaking, Galileo thus did not observe the mountains or valleys of the moon, but observed an interplay of light and shadow which he interpreted, by analogy with terrestrial perspectives, as indicating the presence of lunar mountains and valleys.

SCIENCE AND THEOLOGY: DISTINCT LEVELS OF REALITY

So what other way of conceiving our world might allow us to affirm, preserve, and engage its spatial depths? The natural sciences make extensive use of the notion of 'levels of explanation', an approach which counters inappropriate reductionist tendencies by emphasizing

---

[59] Such as Bittner and Smith, 'A Theory of Granular Partitions'.
[60] For Ninian Smart's influential characterization of these dimensions, see Rennie, 'The View of the Invisible World'. For a more empirical view, see Visala, 'Explaining Religion at Different Levels'.
[61] Hamou, *La mutation du visible*, 63–6; Shea, 'Looking at the Moon as Another Earth'; Spranzi, 'Galileo and the Mountains of the Moon'.

that some explanations might be offered of certain aspects of systems which could not be applied to every aspect of a system, or the system as a whole.[62] The interaction of such levels is complex, and it is becoming increasingly clear that causation exists and operates at multiple levels and in multiple directions within complex biological systems.[63] Whereas those espousing reductionist approaches prematurely argue that the more fundamental levels—such as physical reality—determine the properties and behaviours of higher levels, it is becoming increasingly clear that such 'bottom-up' approaches need to be supplemented by 'top-down' mechanisms.[64] For example, while physical laws underlie all material entities, there nevertheless exist higher-level causal relations that allow the brain to act as a means of creating theories or searching for meaning, without contradicting or overwriting those lower-level physical laws. Consequently, physics does not control the mind, it enables the mind. Such 'top-down' mechanisms cause difficulties for some natural scientists and philosophers in that they are not material causes. However, such difficulties can be mitigated by reconceptualizing the issue in terms of the 'coarse-graining' of nature, or operational networks within nature.[65]

Reality is stratified, and each scientific discipline develops research methods adapted to its specific objects of study. The complexity of the world requires the use of multiple levels of explanation, both within the natural sciences and beyond. As we noted earlier (59–60), the biologist Steven Rose has noted how multiple explanations might be offered as to why a frog jumped into a pond, from the perspectives of physiology, biochemistry, developmental biology, animal behaviour, and evolutionary theory.[66] All these explanations, Rose suggests, are right; they are also different, and combine to offer explanatory depth. Phenomena in the world can be explained at

---

[62] Potochnik, 'Levels of Explanation Reconceived'; Bechtel, 'Levels of Description and Explanation in Cognitive Science'. For a different way of framing such levels, see Rueger, 'Perspectival Models and Theory Unification', 590–2.

[63] Fazekas and Kertész, 'Causation at Different Levels'.

[64] See the evidence assembled in Ellis, *How Can Physics Underlie the Mind?*, 1–28, 133–209.

[65] Bechtel, 'Explicating Top-Down Causation Using Networks and Dynamics'; Flack, 'Coarse-graining as a Downward Causation Mechanism'.

[66] Rose, 'The Biology of the Future and the Future of Biology', 128–9.

several levels, and—despite the protests of those who wish to assert the intellectual hegemony of their own research fields—none can be regarded as definitive, comprehensive, or normative.

The form of 'critical realism' developed by the philosopher and social scientist Roy Bhaskar provides a conceptual framework that affirms the ontological unity of reality, while recognizing that this unity expresses itself at different levels, each demanding a form of engagement which is determined by the distinctive identity of the area of reality under investigation.[67] Bhaskar's account of critical realism—which he earlier described as 'Transcendental Realism' or 'Critical Naturalism'—allowed the active exploration of *social* realities, thus opening up a rich conceptual toolbox for engaging the multiple levels of religious belief, practice, and communities.

This form of critical realism insists that the world must be regarded as differentiated and stratified. Each individual science deals with a different stratum of this reality, which in turn obliges it to develop and use methods of investigation adapted and appropriate to this stratum. Stratum *B* might be grounded in, and emerge from, Stratum *A*. Yet despite this relation of origin, the same methods of investigation cannot be used in dealing with these two different strata. These methods must be established a posteriori, through an engagement with each of these strata of reality. For the purposes of our discussion, we can leave open the question of whether this is simply a heuristic device for conceiving the relation of the sciences in a visual form, or whether we can actually speak of such layers within the overall structure of reality.

On this approach, the three representative sciences of physics, chemistry, and biology can be seen to exist as layers: physics is fundamental; chemistry builds upon physics, while extending it in ways which could not necessarily be predicted on the basis of physics alone; while biology builds further on chemistry, while extending both physics and chemistry in ways that could not have been predicted from a knowledge of those lower levels. Each level is to be regarded as distinct, thus demanding its own method of investigation and representation which is adapted to its structures and forms,

---

[67] For an overview, see Collier, *Critical Realism*.

rather than having some methodology developed for another purpose and application be imposed upon it.

There is an obvious theological counterpart here. Thomas F. Torrance argued that all intellectual disciplines or sciences are under an intrinsic obligation to give an account of reality 'according to its distinct nature'.[68] For Torrance, this means that both scientists and theologians are under an obligation to 'think only in accordance with the nature of the given'.[69] The object which is to be investigated must be allowed a voice in this process of enquiry. The distinctive characteristic of a 'science' is to give an accurate and objective account of things in a manner that is appropriate to the reality being investigated. Both theology and the natural sciences are thus to be seen as a posteriori activities which respond to 'the given' rather than as a priori speculation based on philosophical first principles. In the case of the natural sciences, this 'given' is the world of nature; in the case of theological science, it is God's self-revelation in Christ.

Bhaskar also offers a critical realist account of the relation of the natural and social sciences which affirms their methodological commonalities, while respecting their distinctions, particularly when these arise on account of their objects of investigation.[70]

> Naturalism holds that it is possible to give an account of science under which the proper and more or less specific methods of both the natural and social sciences can fall. But it does not deny that there are significant differences in these methods, grounded in real differences in their subject-matters and in the relationships in which these sciences stand to them . . . It is the nature of the object that determines the form of its possible science.

We see here a clear recognition of each science being characterized by the nature of its object, and being obligated to respond to that object in a manner which is appropriate to its distinctive nature: ontology determines epistemology. If we have 'a conception of the world as stratified and differentiated', the nature of any specific object determines both the *manner* in which it is to be known, and the *extent* to

---

[68] Torrance, *Theological Science*, 10. See further Myers, 'The Stratification of Knowledge in the Thought of T. F. Torrance'.

[69] Torrance, *Theology in Reconstruction*, 9.

[70] Bhaskar, *The Possibility of Naturalism*, 3.

Mapping Human Reason 69

which it can be known. There is thus no *mathesis universalis*, no universal methodology for investigating everything, such as that proposed during the Enlightenment, and echoed by later writers such as Heinrich Scholz.[71]

Bhaskar insists that each stratum of reality—whether physical, biological, or social—is to be seen as 'real', and capable of investigation using means appropriate to its distinctive identity. An example—not used by Bhaskar himself—will help make this point clearer. Consider the concept of 'disability'. It is obvious that it is a complex notion, with multiple levels. Recognizing this complexity, the World Health Organization developed the 'International Classification of Functioning, Disability and Health' (ICF), which recognizes four constitutive levels or strata of 'disability':[72]

1. *Pathology*, in which abnormalities arise in the structure of function of a human organ or organ system.
2. *Impairment*, in which abnormalities or changes arise in the structure or function of the whole human body.
3. *Activity*, in which abnormalities, changes, or restrictions arise in the interaction between a person and their environment or physical context.
4. *Participation*, in which changes, limitations, or abnormalities arise in the position of the person in their social context or environment.

The model is important in creating a framework of understanding which identifies and remedies the inadequacies of reductionist accounts of biosocial phenomena such as disability,[73] or illnesses with social consequences, such as rheumatoid arthritis.[74] A medical model views disability exclusively as a problem of the individual person, caused directly by disease, trauma, or some other health condition, which calls for medical treatment or intervention to correct the problem in the individual. By contrast, a social model of disability

---

[71] McGrath, 'Theologie als Mathesis Universalis?'.
[72] World Health Organization, *International Classification of Functioning, Disability and Health*.
[73] Üstün et al., 'The International Classification of Functioning, Disability and Health'.
[74] Fransen et al., 'The ICIDH-2 as a Framework for the Assessment of Functioning and Disability in Rheumatoid Arthritis'. (ICIDH-2 was a predecessor of ICF.)

conceptualizes disability exclusively as a socially created problem and not an attribute of an individual. On the social model, disability requires social action, since it arises from a dysfunctional social environment. The ICF model synthesizes what is true and useful in both medical and social models, without improperly reducing the whole complex notion of disability to one of its aspects. This stratified account of disability allows a better understanding of both the problem and its potential solutions.

This leads into the critically important question: what research methods might be used to investigate disability? For Bhaskar, 'the nature of objects' determines 'their cognitive possibilities for us'.[75] This means that we cannot use the same research methods to investigate each of these four levels of the complex phenomenon we call 'disability'. We must use methods that are adapted and appropriate to each stratum of reality. A brain tumour is a good example of a pathology with important consequences for human functionality and well-being. Positron emission tomography (PET) is highly effective at detecting and locating brain tumours.[76] Yet this research method has no utility whatsoever in the empirical investigation of changes in an individual's cognitive functionality resulting from the growth of a brain tumour, which is better studied using standard tests such as the Wechsler Adult Intelligence Scale (WAIS).

It is important to appreciate that the recognition of stratification does not imply, still less entail, that the properties of higher strata are determined by, or can be predicted on the basis of, the lower strata. It is a commonplace in popular scientific writings to offer arbitrary reductionist accounts of complex phenomena, such as Francis Crick's simplistic overstatement:[77]

You, your joys and your sorrows, your memories and your ambitions, your sense of personal identity and free will, are in fact no more than the behavior of a vast assembly of nerve cells and their associated molecules.

This represents a reductive explanation of human behaviour which arbitrarily terminates at the molecular level, apparently on the basis of

---

[75] Bhaskar, *The Possibility of Naturalism*, 3.
[76] Chen, 'Clinical Applications of PET in Brain Tumors'.
[77] Crick, *The Astonishing Hypothesis*, 3.

the unstated assumption that the properties and status of lower (though in this case not the lowest) levels determine those of the higher. Biological processes are assumed to always be derivable from lower-level data and mechanisms. Yet this fails to take account of top-down causative processes, and the more general point that there now appears to be no 'privileged' level of causation in the first place.[78]

Bhaskar's critical realism allows scientism to be seen as an unjustified imposition of a single research method appropriate for, and developed in relation to, one specific level of reality on to every aspect of the natural and social world.[79] For Bhaskar, the nature of the object determines the form of its possible science; scientism, however, insists that everything must be investigated using the methods of the natural sciences—even when these are not adapted or appropriate for the investigation of certain critical questions, such as issues of meaning or purpose. Scientism denies that there are 'any significant differences in the methods appropriate to studying social and natural objects'.[80] Scientism thus reduces reality to what can be known through the application of one specific research method—often on the basis of the questionable assumption that there is a single 'scientific method'. Epistemology is allowed to determine ontology, in that the use of one specific research method determines what is 'seen'—and hence judged to be real.

Scientism is, on this approach, blind to the existence of levels of reality that cannot be engaged by the methods of the natural sciences—methods, it must be added, which were developed for other purposes. The observation that a specific research method does not disclose any given level is misinterpreted as implying that this level does not exist.

And what of religion? Religion is a complex reality, with multiple levels or strata. One stratum of religion consists of beliefs about God, salvation, and human identity. This stratum would seem to include the process of intellectual reflection on the foundations and interrelations of such beliefs which is usually understood as theology. Another

---

[78] Noble, 'A Theory of Biological Relativity'; Ellis, *How Can Physics Underlie the Mind?*
[79] Bhaskar, *The Possibility of Naturalism*, 2–3.
[80] Bhaskar, *The Possibility of Naturalism*, 2.

stratum relates to certain kinds of behaviour, such as prayer or participation in worship. A third involves certain kinds of emotion or phenomenological experiences that are usually described or categorized (although not necessarily helpfully) as 'spiritual' or 'religious'. Finally, many religions have specific social structures and institutional forms. Each of these strata can be explored using research methods adapted to its specific nature. The disciplines of sociology of religion and the psychology of religion thus focus on two such strata. Yet none of these strata, alone, is enough to define or characterize a 'religion', as this term is generally used.

This has important implications for many aspects of the study of religion, not least attempts to explain the origins and characteristics of religion on evolutionary grounds. Given the complexity and stratified nature of religion, it proves resistant to explanation on evolutionary grounds in its totality. As a result, most evolutionary accounts of the origins of religion have focussed on a single specific aspect of religion amenable to such explanation, and ignored others. This 'fractionation' of religion is problematic, not least because it fails to take account of the interaction of such components within a wider system.

Mounting criticisms of cognitive-evolutionary theories of religion reflect a concern that these generally fail to recognize the complexity of what is conventionally understood by the term 'religion', and the inadequacy of the use of monochromatic theorizing in explaining the existence of polychromatic phenomena.[81] For example, some theories focus on group behaviour such as ritual, arguing for an adaptive function related to the benefits to the group arising from increased cohesion or solidarity. An adaptationist explanation of, for example, belief in God does not plausibly correlate with specific rituals or ethical norms.[82] A strong evolutionary account of the origins of religion would have to account, not merely for its individual aspects, but the manner of their correlation in cultural practice.

This stratified approach to science and religion is capable of preserving and accommodating their complexity and stratification. It recognizes that the natural sciences exist in relationships of interaction and dependency, and that religion is a multi-layered phenomenon

---

[81] For example, see Saler, *Conceptualizing Religion*; idem, 'Theory and Criticism'.
[82] See the analysis in Kirkpatrick, 'Religion Is Not an Adaptation'.

Mapping Human Reason 73

which cannot be reduced to any of its communal, symbolic, narrative, or ideational elements. It is also capable of accommodating the shifting historical and cultural understandings of what each of the terms 'science' and 'religion' designates.[83] One of the more fundamental concerns about Ian Barbour's fourfold taxonomy of relationships between science and religion—conflict, dialogue, independence, and integration—is that it is of severely limited utility in allowing engagement with historical debates, in that these are socially and culturally embedded, often involving the dynamics of institutions (such as the church), cultural associations, and historical memories, rather than the mere relation of ideas.[84] A stratified approach to both science and religion does not displace, but rather complements, perspectival approaches.

Yet in the end, it is not entirely clear how such multiple perspectives and levels are to be woven together, in that such a process involves judgement about what weight is to be attached to each voice, and how seeming inconsistencies are to be addressed. Any 'transversal' approach to rationality,[85] which, recognizing the Enlightenment's failure to provide an adequate defence of its own *Letztebegründung*,[86] seeks to be attentive to many voices and social practices, must make transparent and warranted decisions about how these voices and practices are to be assessed and integrated. In the end, the notion of transversality is fundamentally a heuristic device that creates imaginative space for affirming such multiple approaches, rather than a conceptual algorithm for calibrating their competing claims to authority, or the outcomes of their application.

Yet however imprecise we might find the notions of multiple perspectives and levels, both these approaches nevertheless offer an imaginative framework which allows us to see how multiple

---

[83] Harrison, *The Territories of Science and Religion*.
[84] Cantor and Kenny, 'Barbour's Fourfold Way'.
[85] The best account is Welsch's sprawling treatise *Vernunft*, which merits close study despite its dense style. For two more accessible North American explorations of this theme, see Schrag, *The Resources of Rationality*, 148–79; van Huyssteen, *Alone in the World?*, 20–3. Recent discussion of 'transdisciplinarity' has not drawn on Welsch's analysis, although there are illuminating parallels between them.
[86] See, for example, d'Alembert's appeal to metaphysics as such an ultimate ground of knowledge: Neuser, *Natur und Begriff*, 99–123.

approaches might be held together and correlated, and seen as part of a greater enterprise of securing traction on a complex reality. Human rationality takes the form of a spectrum of practices, developed and adapted to a variety of situations and tasks encountered in the process of production of knowledge. Any attempt to achieve a broader vision of reality than that offered by a single discipline must, however, find some way of holding such insights together in the first place, if a grander vision of reality is to emerge.

3

Social Aspects of Rationality

Tradition and Epistemic Communities

Human beings are not solitary creatures. As Aristotle pointed out, we are communal animals—*politikoi* in the sense that we inhabit a *polis*, a community.[1] An individual's social location is one of a number of factors which help shape perceptions of what beliefs or criteria of judgement might be deemed to be 'reasonable', and thus influence the outcomes of the process of reflection.[2] The manner in which a group of individuals arrives at a collective decision is not the same as that in which an individual develops such a judgement. Groups—whether juries or societies—arrive at their aggregated beliefs in ways that are shaped by social considerations, often involving the collectivizing of reason, which can paradoxically lead to the group's collectively endorsing a conclusion that a majority of the group members individually reject.[3]

The standard textbook presentations of the development of scientific thought or religious ideas typically present individuals as detached geniuses, individuals who single-mindedly brought about a revolution in the way in which we think about ourselves and our universe. To suggest that the situation is rather more complex than this is not in any way to detract from the achievements of Copernicus or Darwin,

---

[1] Kullmann, *Aristoteles und die moderne Wissenschaft*, 334–63.
[2] For the implications of this observation for reflecting on rationality, see Tuomela, *The Philosophy of Sociality*.
[3] Pettit, 'Groups with Minds of Their Own', especially 175–8. More generally, see Rupert, 'Minding One's Cognitive Systems'.

Aquinas or Luther—to name just a few of these individuals. Both the natural sciences and Christian theology are to be seen as communal activities, in which ideas are generated, evaluated, and received according to procedures and norms which emerge within those communities, and are believed to be appropriate for their distinct resources and tasks. In this chapter, we shall consider the implications of critical historical analysis and social epistemology for an understanding of communal rationalities, and how these might illuminate discussions about rationality within both religious and scientific epistemic communities.[4]

COMMUNITIES AND THEIR EPISTEMIC SYSTEMS

It is a matter of empirical observation that communities develop their own distinct 'epistemic systems'—that is, characteristic systems of rules or principles about the conditions under which a given belief may be said to be justified.[5] Close studies of authors of the early modern period have made it clear that they did not subscribe to a *single* valid mechanism of knowledge-production, in that their epistemological views were motivated by multiple sources based on their differing ideological backgrounds.[6] Yet this observation also applies to the communities within which they were embedded, which both informed and sustained the 'plausibility structures' of such epistemological views. As Peter Berger has argued, belief systems—whether religious or secular—are socially constructed and thus require social confirmation and validation through participation in networks of individuals who share these beliefs.[7] A failure to participate in such networks makes the beliefs easier to doubt, especially in a society characterized by pluralistic beliefs, many of which may contradict or compete with those of the individual. Epistemic communities offer plausibility structures which thus reinforce and affirm those beliefs. Cognitive science may help explain why certain religious

---

[4] Zollman, 'The Communication Structure of Epistemic Communities'.
[5] Boghossian, *Fear of Knowledge*, 58–80.
[6] Dear, 'Reason and Common Culture in Early Modern Natural Philosophy'.
[7] Berger, *A Far Glory*, 127–8.

beliefs seem natural to individuals; the communal context nevertheless remains important in consolidating these beliefs and giving them specific form.[8]

Recent revisionist accounts of the points of divergence between the English philosophers John Locke and John Sergeant during the 1690s have highlighted the importance of social factors in the production of knowledge.[9] According to Sergeant, an epistemology that attempts to explore the world based only on the knowing subject (and some specific capacities of the mind) is unacceptably individualistic, in that it neglects the community within which the thinking subject was embedded, and whose traditions and accumulated judgements shaped such a thinker's view. A thinker's communitarian context was to be seen as a potential source of knowledge-acquisition, whether this was to be seen as a positive or negative influence. A similar concern was expressed a century later by Johann Georg Hamman, who questioned whether Kant's understanding of human reasoning was sufficiently attentive to the consequences of the thinking subject being embedded within a specific cultural context.[10] Both Sergeant and Hamman, though in different ways, expressed a fundamental concern that the dominant approach to questions of knowledge and justified belief was heavily individualistic in focus, and thus offered a distorted picture of the human epistemic situation.

This does not require us to conclude that the variety of observed rationalities entails some form of relativism; it does, however, alert us to the importance of social factors in shaping human judgements, and the difficulties that can arise from this. It also illuminates the observation that a single body of evidence can be interpreted in multiple ways, without implying that only one such interpretation is 'rational'. Although this issue is familiar from the natural sciences—consider, for example, the debate about whether there is a single universe or a vast ensemble of universes, in which the same observations are

---

[8] Gervais, Willard, Norenzayan, and Henrich, 'The Cultural Transmission of Faith'; Luhrmann, Nusbaum, and Thisted, 'The Absorption Hypothesis'.

[9] Levitin, 'Reconsidering John Sergeant's Attacks on Locke's *Essay*'; Henry, 'Testimony and Empiricism'.

[10] Bayer, *Vernunft ist Sprache*, 252–4; Hempelmann, 'Keine ewigen Wahrheiten als unaufhörliche zeitliche'. For the general issue, see Sandel, *Liberalism and the Limits of Justice*.

interpreted in quite different manners[11]—it is also a commonplace in any collective interpretation of evidence. When faced with an identical body of evidence, individual members of a group—such as a jury—may assess the evidence in different manners, without being irrational in doing so.[12] The collective decision reached by such a group often masks the diversity of doxastic attitudes within it, and the processes of negotiation by which these are resolved.[13]

It is also important to note that 'rationality' is often rhetorically equated with the groupthink of the cultural establishment, or of groups with a sense of cultural entitlement. It can be suggested that societal norms of rationality both perpetrate and perpetuate forms of epistemic injustice, in that these often distribute such rational credibility unjustly, assigning it to preferred social groups, such as the privileged or powerful.[14] When a group takes on or is given an epistemic task, its performance is partly determined by 'aggregation procedures'—that is, its mechanisms for consolidating the group members' individual beliefs or judgements into corresponding collective beliefs or judgements endorsed by the group as a whole.[15]

A modern Western nation state consists of a variety of epistemic communities jostling for social, political, and intellectual acceptance, and occasionally hegemony. In the case of the United States of America, this has led to a fractured society of 'beliefs and intuitions resting on tradition-dependent values that cannot be empirically proven or fully justified by forms of rationality external to those traditions.'[16] Some, disturbed by such observations, respond by reasserting the hegemony of one specific conception of rationality—such as that of the European 'Age of Reason', or those associated with the 'hard sciences', such as physics—while ridiculing those who suggest that there is no independent vantage point, no privileged seat of judgement, by which rival traditions of enquiry and adjudication can be assessed.

---

[11] See Carr, ed., *Universe or Multiverse?*
[12] Rosen, 'Nominalism, Naturalism, Philosophical Relativism', 71.
[13] Pettit, 'When to Defer to Majority Testimony—and When Not'.
[14] See the careful analysis in Fricker, 'Rational Authority and Social Power'.
[15] List, 'Group Knowledge and Group Rationality'; List and Pettit, 'Aggregating Sets of Judgments'. See also Goldman, 'Group Knowledge versus Group Rationality'.
[16] Inazu, *Confident Pluralism*, 88.

Such epistemic anxiety lay behind the convening of the World Congress of the International Academy of Humanism (2005) by the leading secular humanist Paul Kurtz, who declared that there was an urgent and pressing need for a 'New Enlightenment'.[17] After listing the Enlightenment's achievements with an enthusiasm that was untroubled by the inconveniences of historical accuracy, Kurtz warned that there had been 'a massive retreat from Enlightenment ideals in recent years, a return to pre-modern mythologies', including a receptivity towards a 'vulgar post-modernist cacophony of Heideggerian-Derridian mush'. The promised 'New Enlightenment' would be based on 'scientific inquiry and philosophical rationality', offering what its critics could only see as a fundamentally reversionary hermeneutic tendency in cultural development, based on nostalgia for a highly idealized and sanitized 'Enlightenment'.[18]

Yet Kurtz's enthusiastic promissory note was not succeeded by its intended payload, largely because of an ideologically motivated disinclination to engage either the critical historical scholarship of the origins and development of the Enlightenment, or the intellectual objections raised to what might loosely be described as the 'Enlightenment Project' by Heidegger, the later Wittgenstein, or Gadamer.[19] Human reasoning operates within an already given network of assumptions and motives, so that any given conception of rationality will always be relative to some informing context. The rational justification of our most fundamental values is potentially a circular process, in that those values and beliefs constitute the context within which our understanding of rationality functions.[20]

---

[17] Kurtz, 'Re-enchantment: A New Enlightenment'.
[18] See also Hitchen's similar plea for a 'New Enlightenment': Hitchens, *God Is Not Great*, 277–83.
[19] It is instructive to read Kurtz's comments in the light of the proceedings of the 23rd International Wittgenstein-Symposium (2000), collected in Brogaard and Smith, eds, *Rationalität und Irrationalität*.
[20] Cf. Strawson, *Skepticism and Naturalism*, 39: 'We have an original non-rational commitment which sets the bounds within which, or the stage upon which, reason can effectively operate, and within which the question of the rationality or irrationality, justification or lack of justification, of this or that particular judgment or belief can come up.'

RATIONALITY, COMMUNITY, AND TRADITION

In his *Genealogy of Morals*, Nietzsche argued that our vision of the world is embedded, local, and perspectival. We are socially embedded creatures, and cannot escape the particularities of our social and cultural contexts, which shape our assumptions and outlooks.[21] Thomas Nagel thus argues that we cannot escape the condition of seeing the world from the particular point at which we have been inserted within it.[22] No matter how much we may aspire to conditions of absolute cultural detachment and objectivity, we are forced to settle for a view that is 'incurably open to bias and limitation'.[23]

Many would resist such a conclusion, hoping to find some vantage point which is exempt from the contingencies of the rational thinker's location within the historical process. The core rational virtues of objectivity and neutrality seem to require a capacity to stand outside history. Yet the quest for culturally invariant norms of human reasoning, such as those aspired to within sections of the Enlightenment, has encountered fundamental difficulties, both diachronically and synchronically. Both Baruch Spinoza and Moses Mendelssohn—two leading figures of the Enlightenment—found themselves negotiating a delicate boundary between the *universality* of their vision of human reasoning and the *specificity* of their Jewish identities.[24] They were, so to speak, citizens of two intellectual territories, members of two social communities with potentially different and divergent epistemic norms.

The issue here is partly that of the cultural embeddedness of human thought, and the desire to escape from the limits imposed by our historicity. Some have turned to logic and mathematics as intellectual paradigms for modes of reasoning which appear to be culturally invariant. As Stephen Toulmin pointed out, the attraction of pure mathematics to the 'Age of Reason' lay partly in the fact that it

---

[21] Heelan, 'Nietzsche's Perspectivalism'; Emden, *Nietzsche's Naturalism*, 184–203.
[22] Nagel, *The View from Nowhere*, 67–89.
[23] Weinstein, 'The View from Somewhere', 85.
[24] A point brought out particularly clearly by Goetschel, *Spinoza's Modernity*, 3–20, 91–2.

was seen to be the only intellectual activity whose problems and solutions were 'above time',[25] perhaps offering hope that philosophy and ethics might achieve a similar status. Yet other intellectual disciplines—including both the natural sciences and religion—are embedded within their historical and cultural contexts, with potentially indeterminate yet significant implications for their patterns of reasoning.

These concerns have led some to conclude that we must speak of the 'rationality of traditions', recognizing that communities develop, propagate, and sustain specific implementations of rationality. On this approach, a rationality is embedded within the values and practices of a community. One of the most important of these exploratory projects is due to Alasdair MacIntyre,[26] who argued that a study of the history of ethical thought demonstrated *empirically* that a neutral tradition-independent ground from which a verdict may be passed upon the rival claims of conflicting traditions in respect of practical rationality and of justice had yet to be discovered, and suggested *theoretically* that rationality was itself constituted and mediated through traditions.[27] The history of ideas simply did not lend support to the notional of culturally invariant moralities or rationalities; the question of how this observation is to be interpreted remains open, although MacIntyre's approach to the question has gained considerable traction, not least because it is able to accommodate the observation that distinct rationalities arise within individual epistemic communities.

The concept of an 'epistemic community' was developed to help make sense of the complex interaction between professional communities, their values, their rationalities, and their formulation of policies. Such communities hold a shared set of normative and principled beliefs, which provide a value-based rationale for the social action of community members, which are derived and justified on the basis of shared notions of validity—that is, 'intersubjective, internally

---

[25] Toulmin, *The Uses of Argument*, 118.
[26] For recent accounts and critiques, see Herdt, 'Alasdair Macintyre's "Rationality of Traditions" and Tradition-Transcendental Standards of Justification'; Nicholas, *Reason, Tradition, and the Good*; Seipel, 'Tradition-Constituted Inquiry and the Problem of Tradition-Inherence'.
[27] MacIntyre, *Whose Justice? Which Rationality?*, 9.

defined criteria for weighing and validating knowledge in the domain of their expertise'.[28] Although there are clear parallels with MacIntyre's approach, the concept of an 'epistemic community' tends to emphasize procedural rationalities, taking cues from the work of Herbert A. Simon, who received the 1978 Nobel Prize in Economic Sciences for his pioneering work on decision-making processes in economic organizations, rather than that of MacIntyre.[29]

MacIntyre himself was clear that there was no a priori reason to suppose that a 'neutral tradition-independent ground' of judgement did not exist;[30] his point was rather that the historical and present failure to achieve fundamental agreement on such a ground was an a posteriori reason for supposing that it did not exist. Although MacIntyre is perhaps less clear on how we are to define the critically important concept of 'tradition' than many would like,[31] he in effect adopts an instrumentalist reading of the notion that he appears to consider adequate for his purposes. MacIntyre does not understand tradition in a conservative sense, as of a fixed body of teachings or beliefs which are mechanically transmitted over time, but as a community of discourse within which disagreement elicits intellectual advance.[32]

> There is no other way to engage in the formulation, elaboration, rational justification,and criticism of accounts of practical rationality and justice except from within some one particular tradition in conversation, cooperation, and conflict with those who inhabit the same tradition. There is no standing ground, no place for enquiry, no way to engage in the practices of advancing, evaluating, accepting, and rejecting reasoned argument apart from that which is provided by some particular tradition or other.

This line of analysis leads MacIntyre to draw three main conclusions:[33]

---

[28] Haas, *Epistemic Communities, Constructivism, and International Environmental Politics*, 5. The term 'epistemic community' is now generally used in a more general sense to designate any group of experts giving policy advice.

[29] See especially Simon, 'Rationality as a Process and as a Product of Thought'; idem, *Reason in Human Affairs*.

[30] MacIntyre, *Whose Justice? Which Rationality?*, 346.

[31] A concern expressed in Allen, 'Macintyre's Traditionalism'. For a fuller discussion, see Trenery, *Alasdair Macintyre, George Lindbeck, and the Nature of Tradition*.

[32] MacIntyre, *Whose Justice? Which Rationality?*, 350.

[33] Seipel, 'In Defense of the Rationality of Traditions', 258.

1. Rationality is dependent on the resources of traditions;
2. There is no neutral, tradition-independent way in which to assess the epistemic status of a theory or belief; and
3. One tradition can rationally defeat another by argument.

Many have expressed concern about the consequences of this approach, which are taken to be indicative of a more fundamental flaw in its foundations. MacIntyre, it is argued, makes claims about the possibility of rational evaluation across traditions which are irreconcilable with his conception of the tradition-dependent nature of rationality. This suggests either that his theory of the rationality of traditions has a tradition-independent basis (in which case his conception of rationality is false), or that the theory is merely to be regarded as justified within and for a particular tradition (and therefore fails to refute relativism).[34] On this reading of MacIntyre, his theory of the rationality of traditions must be universally valid if he is to be able to refute relativism; yet any suggestion of its universal validity is inconsistent with his emphasis on the dependence of rationality on specific traditions. Such a theory of rationality would fall into the category of 'tradition-transcendental', in that it would be neither limited to a specific tradition nor justified in a tradition-independent sense, providing a rational standard that is 'not limited to a particular tradition, even if only at a general, procedural level'.[35]

Yet our concern here is not with defending the positive notion of 'tradition-dependent rationality' which MacIntyre affirms, but with noting his empirical argument that understandings of what is 'rational' are to be located within some particular, historical, and contingent tradition of theoretical enquiry that is socially embodied. If this is so, we are obliged to speak and think in terms of the actual use of and appeal to *multiple* rationalities. This leaves open the question of the relationship of these rationalities, and whether they might be seen as specific yet *partial* historical or cultural implementations of some grander meta-tradition.[36] Yet it suggests that the idea

---

[34] Seipel, 'In Defense of the Rationality of Traditions', especially 257–8.
[35] Herdt, 'Alasdair Macintyre's "Rationality of Traditions" and Tradition-Transcendental Standards of Justification', 535.
[36] I explore this idea in McGrath, 'The Rationality of Faith'.

that there is some overarching concept of rationality, appropriate to all traditions and communities of enquiry, is not justified in the light of the evidence about how such communities and traditions actually practise their intellectual and reflective tasks.

RATIONALITY AND DOMINANT CULTURAL METANARRATIVES

The origin of MacIntyre's notion of the 'rationality of traditions' lay in his historical analysis of understandings of ethical and rational norms of the Enlightenment project, which convinced him that its legacy was an ideal of rational justification which proved impossible to attain in practice.[37] McIntyre's approach could be seen in terms of the historicization of rationality—the recognition that a good theory of rationality should somehow fit the history of human rational enterprises, such as the natural sciences. Critical historical investigation raises questions about any assumption that thinkers of the past shared assumptions about the definition or implementation of rationality.[38] Core assumptions of the past, such as that females were less rational than males, have been left behind us;[39] yet the fact that these assumptions were made, accepted, and regarded as rational must heighten our sensitivity towards the social location of discussion of human rationality. A belief that one culture saw as eminently rational might be seen by another as irrational, even as a sign of mental illness.[40]

In his *Secular Age*, Charles Taylor considers the question of why a belief that was regarded as eminently reasonable in 1500 might be seen as unreasonable or problematic in 2000. After all, if rationality is a cultural constant, it might be anticipated that a belief could be

[37] MacIntyre, *Whose Justice? Which Rationality?*, 6.
[38] A good example is Thomas Kuhn's challenge to older assumptions about scientific rationality: see Friedman, 'Kuhn, and the Rationality of Science'. Cf. Webel, *The Politics of Rationality*, passim.
[39] For the wider issues, see Jones, 'Rationality and Gender'; Heikes, *Rationality and Feminist Philosophy*.
[40] On the social construction of 'madness', see Walker, 'The Social Construction of Mental Illness'. The classic study of social irrationality remains Foucault, *Folie et déraison*.

deemed rational once and for all. Yet it is clear that some notions once regarded as entirely reasonable—such as the idea that God acts in the world—are now seen as problematic.[41] Of course, examples also abound of ideas that were once seen as problematic which are now seen as entirely reasonable. Newton and his late seventeenth-century contemporaries found the notion of gravity deeply counterintuitive, in that it involved the apparently inexplicable notion of 'action at a distance'. There was simply no conceptual space for *actio in distans* within the corpuscularist worldview of Newton's age.[42] Today, however, this notion is seen as quite unproblematic.[43] Perhaps the tipping point was Helmholtz's 1847 declaration that any attempt to render the universe intelligible depended on forces of attraction or repulsion, the intensity of which depended upon distance.[44] What was once seen as a potential irrationality thus came to be recognized as a fundamental principle of cosmic intelligibility.

As we noted earlier (25), however, three elements can be discerned within any analysis of what a culture deems to be reasonable: human cognitive processes; the dominant presuppositions of the culture within which a thinker is located; and the evidence available to that culture. Taylor's rich analysis of cultural change in the West highlights the importance of cultural metanarratives in determining what are considered to be reasonable beliefs and values. It illuminates some of the apparent contradictions and paradoxes of the Enlightenment, such as the white racial framing mechanisms of the intellectual elite of Enlightenment France,[45] and the curious decision of the founders of the United States of America to preserve slavery and social inequality after winning their political freedom from Britain. Even David Hume, whose opposition to slavery during the 1770s put him ahead of his age, held views that would now be seen as racist, evident in his remark that he was 'apt to suspect the negroes to be naturally inferior to the whites'.[46]

---

[41] See the analysis in McGrath, 'Hesitations About Special Divine Action'.
[42] Ducheyne, 'Newton on Action at a Distance'.
[43] Dijksterhuis, *De Mechanisering van het Wereldbeeld*, 512–13.
[44] von Helmholtz, *Über die Erhaltung der Kraft*, 17.
[45] On which see Curran, *The Anatomy of Blackness*, 117–66, 216–24.
[46] Hume, *Essays, Moral, Political, and Literary*, 208. This 1777 statement is much less offensive than an earlier version of 1753, which was widely cited in racist literature. For comment and analysis, see Immerwahr, 'Hume's Revised Racism'.

Taylor's *Secular Age* charts the rise to social dominance of a cluster of modern attitudes which he designates 'The Immanent Frame'. This cultural metanarrative weaves together a number of themes, including the disenchantment of the world,[47] an understanding of nature as an impersonal order, the rise of an 'exclusive humanism', and an ethic which is framed primarily in terms of discipline, rules, and norms. This 'exclusive humanism' advocates a view of human flourishing which denies or suppresses any notion of a transcendent source of morality, such as God or the Tao, and which refuses to recognize any good beyond this life and world. For Taylor, contemporary understandings of human flourishing, the natural order, the moral life, and nature are thus framed in a self-sufficient, naturalistic, and immanent manner.

The culture of our 'secular age' now makes a sharp distinction between the natural and the supernatural, the human and the divine, so that making sense of the world around us now seems to be possible in terms of this world alone. Nature became emptied of the spirits, signs, and cosmic purposes that once seemed a fact of everyday experience. It came to be conceived fundamentally as an impersonal order of matter and force, governed by causal laws. There has thus been a marked shift to 'Closed World Structures' that implicitly accept the 'immanent frame' as normative, evident in the fact that most people no longer see natural events as acts of God.[48] Nature has become reduced to the predictable and quantifiable. For Taylor, this means that the dominant cultural narrative leaves no place for the 'vertical' or 'transcendent', but in one way or another closes these off, renders them inaccessible, or even unthinkable. 'Closed World Structures' now function as unchallenged axioms in Western culture.

Taylor's analysis, though not without its difficulties, offers some illuminating insights into how Western culture became secular, and what this implies. Yet the importance of his discussion for our purposes lies in his identification of the importance of the grand narratives or conceptual frameworks which capture the imagination

---

[47] For an alternative reading of this cultural trend, see Josephson-Storm, *The Myth of Disenchantment*, 269–300.

[48] Taylor, 'Geschlossene Weltstrukture in der Moderne'.

of a culture, and thus shape its notions of rationality.[49] These 'social imaginaries' are, however, cultural artefacts, not empirical realities; social constructions, not observational givens; historical contingencies, not rational necessities. They are not predictable; they arose in certain historical contexts, and will change over time—as will any notions of rationality that are shaped by them.

This analysis reinforces the growing consensus that the notion of a 'universal rationality' is a historical fiction, in that it is clear that different cultures in the past developed and maintained quite distinct understandings of rational processes and criteria of judgement, located within their own 'social imaginaries'. But what of the future? Will the process of globalization lead to cultural homogenization, so that one single cultural metanarrative (or family of related metanarratives) might achieve dominance?

Some would argue that the Enlightenment's 'cosmopolitan, universalizing vision of the human world'[50] ultimately leads to a vision of civilization in which questions of justice can be applied and upheld at a global level by a shared conception of rationality. Some such vision certainly lay behind the colonial programmes of education developed by European powers during the nineteenth century, which often sought to impose their own intellectual and cultural values on what they generally regarded as 'primitive' nations, not necessarily because of a desire to control minds, but often on the basis of the assumption that such ways of thinking were self-evidently right, and would thus serve as the basis of a cosmopolitan culture.[51] Yet why should such a cosmopolitanism be allowed to be grounded on ideals that were developed centuries ago in a specific historical and cultural location—namely, eighteenth-century Western Europe?[52] This vision of a global 'civilizing process' exudes the condescending cultural values of a colonial age, making it vulnerable to those who demand intellectual and cultural autonomy, rather

---

[49] McKenzie, *Interpreting Charles Taylor's Social Theory on Religion and Secularization*, 90–3.
[50] Pagden, *The Enlightenment and Why It Still Matters*, 323.
[51] Ghosh, 'English in Taste, in Opinions, in Words and Intellect'; Hall, 'Making Colonial Subjects'.
[52] A point raised at multiple points by African scholars in Hountondji, ed., *La rationalité, une ou plurielle?*

than to have someone else's rational and cultural norms imposed upon them.

It can argued that the concepts of rationality associated with the European Enlightenment gained global traction for a number of reasons, one of which is the colonial enterprise itself, in which European cultural and rational norms were imposed upon, or privileged within, indigenous cultures.[53] Intellectual colonization was an integral aspect of a wider cultural project, in which Western nations sought to acquire territory, resources, and influence;[54] yet although it may not have been a primary objective, it nevertheless often led to the suppression or marginalization of native rationalities.

Yet there are now questions about whether globalization has faltered, or fundamentally changed its character. Calls for 'epistemological decolonization' reflect both an awareness of the multiplicity of potential modes of reasoning, and a reaction against the imposition of Western understandings of rationality.[55] The notion of a 'local rationality'—which has clear affinities with MacIntyre's 'rationality of traditions'—has acquired political traction within the narrative of decolonization, and raised questions about the future shape and prospects of a 'universal rationality', seen by some as integral to the globalization project.[56] It is, for example, perfectly clear that there are multiple visions of globalization,[57] and this fact raises questions about how these might be resolved and implemented.

So what are the implications of these broad considerations for an understanding of rationality in the natural sciences or religion? The point here is that the social organization of both the scientific and the religious communities has an impact on the knowledge produced by each community, even though there are divergences within the literature concerning which features of that social organization are

---

[53] See, for example, Beck, Bonss, and Lau, 'The Theory of Reflexive Modernization'; Lee, 'In Search of Second Modernity'. For a wider survey of the colonial enterprise and its motivations, see Abernathy, *The Dynamics of Global Dominance*.

[54] For some ideologies of colonialism, see Pyenson, *Empire of Reason*, 1–17; MacMillan, 'Benign and Benevolent Conquest?'

[55] See, for example, Mignolo, *Desobediencia epistémica*, 9–17.

[56] For the notion of a 'local rationality', see Townley, *Reason's Neglect*, 133–45; Cox and Nilsen, 'What Would a Marxist Theory of Local Movements Look Like?', 74–7.

[57] See the contributions to Rupert, ed., *Ideologies of Globalization*.

of particular importance, and the manner in which they are reflected in the theories and models accepted by a given community. Two studies are of particular relevance here: Karin Knorr-Cetina's study of a plant science laboratory at Berkeley, and Bruno Latour and Steven Woolgar's study of the neuroendocrinology laboratory at the Salk Institute for Biological Studies in San Diego.[58] These works suggested that purely philosophical analyses of core concepts such as rationality, evidence, and knowledge were generally of little relevance to understanding how scientific knowledge was actually acquired and tested.[59] Scientific rationality was to be investigated through scientific practice, and relates primarily to a research community, rather than the individuals which constitute this community. And, like any epistemic community, scientific research groups experience tensions over the resolution of differences, in which social status or reputational standing within the group often proves important in the decision-making process.[60]

SCIENCE AND RELIGION: REFLECTIONS ON THE COMMUNAL ASPECTS OF KNOWLEDGE

In this chapter, we have focussed on knowledge production within communities, and reflected on some of the factors that have shaped this knowledge. While popular accounts of scientific discovery tend to portray scientists as solitary geniuses solving significant problems in splendid isolation, the social development of science has increasingly moved away from individuals towards large groups. The rise of large research groups bringing together individuals with different bodies of expertise to a common research project—now widely known as 'Big

---

[58] Knorr-Cetina, *The Manufacture of Knowledge*; Latour and Woolgar, *Laboratory Life*. See further Knorr-Cetina, *Epistemic Cultures*; Solomon, *Social Empiricism*, 117–35.

[59] For reflections on such approaches, see Gooday, 'Placing or Replacing the Laboratory in the History of Science?'

[60] For this point in relation to legal epistemic communities, see van Waarden and Drahos, 'Courts and (Epistemic) Communities in the Convergence of Competition Policies'.

Science'—has introduced significant new dynamics into the production of scientific knowledge. The research group increasingly functions as an interdisciplinary community, bringing an array of skills together for the purpose of investigating a specific topic, or solving a particular problem. Before the First World War, most research took place in small laboratories and often involved small groups of researchers with specialist knowledge in the same discipline.[61]

The Second World War marked a significant change. The 'Manhattan Project' (1942–7) set up to produce the first atomic bomb brought together a large team of scientists with multiple skills and expertise, working in subgroups to deal with specific aspects of the design and production of such a weapon. No one individual had the necessary knowledge to deal with the theoretical questions attending the process of nuclear fission, or the technical experience to develop this into a serviceable weapon. As a result, individuals came to depend on each other for knowledge, in a process which might be described as 'epistemic dependence' or 'social epistemic interdependence'.[62] In effect, it is the research group as a whole which is the 'knower', rather than its individual members. The group makes up for the inadequacies of individuals, bringing together a shared knowledge that was not accessible to any one individual.

As research tasks become increasingly complex, such research groups become correspondingly significant. If I might be allowed a personal reflection, I was a member from 1974 to 1977 of the Oxford Enzyme Group,[63] a research group that was formed at Oxford University in 1969 specifically to create a co-located multidisciplinary team with skills in physical science, biology, and biochemistry to investigate aspects of enzyme structure and function. The research culture that emerged possessed a corporate knowledge of such techniques as nuclear magnetic resonance spectroscopy (NMR), the production of certain enzymes such as lysozyme, and the mathematical modelling of protein structures.

---

[61] Nye, *Before Big Science*.
[62] Hardwig, 'Epistemic Dependence'; Schmitt, 'On the Road to Social Epistemic Interdependence'. Note also the three approaches to epistemic dependence set out in McMyler, *Testimony, Trust, and Authority*, 77–94.
[63] On this group, see Williams, Chapman, and Rowlinson, *Chemistry at Oxford*, 259–61.

Yet such decentralized 'radically collaborative research' involving investigators with different forms of expertise and knowledge raises important questions about the methodological norms of such research, in that researchers from different disciplines bring their own understandings of methods and rationalities to the group as a whole.[64] 'Massively epistemically distributed' scientific collaborations of this kind raise important questions about which methodological practices and rational norms are to be respected, when the individual collaborators come from different epistemic communities with potentially conflicting norms. These divergences often make it 'impossible to assume a coherent, unified set of methods and standards governing the study'. In practice, however, local differences in epistemic standards appear to cancel one another out to yield generally reliable results.

The situation in theology is somewhat different. Like the natural scientist, the theologian increasingly is not an isolated thinker, but someone embedded within epistemic communities—such as an academic faculty, and a community of worship. The theologian thus has the possibility of interacting with, for example, historians of Christian thought and New Testament scholars, thus creating a culture of epistemic dependency on the one hand, while at the same time making possible a broader intellectual engagement than is possible for a single scholar. Yet the theologian is also aware of standing in continuity with (if not in the presence of) earlier generations of theologians who have wrestled with the same questions, and whose works serve as resources for reflection on those themes.

Yet perhaps the more important is a community of worship, which is primarily about the instantiation of the *practice* of theology, and thus shapes theological reflection and articulation.[65] Prosper of Aquitaine's remark that *legem credendi lex statuat supplicandi* is open to multiple interpretations;[66] it nevertheless affirms the inseparability—but not the identity—of the communal practices of prayer and worship and the domain of theological interpretation. The Christian epistemic

---

[64] Winsberg, Huebner, and Kukla, 'Accountability and Values in Radically Collaborative Research'.

[65] For a range of interpretations of this point, see Johnson, *Praying and Believing in Early Christianity*; Knop, *Ecclesia Orans*; Pickstock, *After Writing*.

[66] De Clerck, 'Lex Orandi—Lex Credendi'.

community is one that is grounded and shaped by the narrative of Jesus Christ, and thus expresses itself both in worship and reflection, as two natural and interconnected outcomes of an encounter with the mystery of faith.

But what of science and religion? Does this term designate a coherent and distinct field of study and research? Or is it merely an interface between two well-established and mature fields, marked by the potential vulnerability of all interdisciplinary undertakings, namely a propensity to attract those with little knowledge of either discipline, while at the same time possessing the capacity to enrich both through the forging of new connections? At present, the indications are that this is best seen as an interdisciplinary enterprise with aspirations to become a field in its own right. Yet the potential for development is clearly present, with journals, academic chairs, scholarly societies, and regular conferences dedicated to this broad area.

Yet if the field of science and religion is to develop in this way, the epistemological and methodological issues raised by 'radically collaborative research' can hardly be avoided. As has been argued throughout this work, disciplines operate with distinct and sometimes divergent understandings of rationality, raising the question of how these can be correlated, if they cannot be conflated. We shall return to this question in the concluding chapter of this work.

II

Rationality in Science and Theology: A Critical Engagement

II

4

Rational Virtues and the Problem of Theory Choice

C. S. Lewis is one of many to argue that the capacity to dissociate the process of reflection from vested interests and personal bias is an integral aspect of good scholarship and intellectual enquiry. 'In the moral sphere, every act of justice or charity involves putting ourselves in the other person's place and thus transcending our own competitive particularity. In coming to understand anything we are rejecting the facts as they are for us in favour of the facts as they are.'[1] So what rational virtues need to be cultivated and applied in order to transcend our own limitations and prejudices?

One important answer, which was considered in the previous chapter, is to affirm the importance of communal, rather than individual, assessments of both the available evidence and the best interpretation of this evidence. It might be hoped that these communal insights, whether synchronic or diachronic, help filter out both individual bias and intellectual vested interests. Our concern in this chapter, however, is with identifying the epistemic virtues that might help ascertain the best interpretation of our observations and experience. Yet before turning to consider the problem of theory choice in science and theology, we must give consideration to what is meant by 'theory', and how this impacts on the rational virtues to be valued and deployed in its pursuit.

Perhaps it is understandable that many consider that the process by which a theory is confirmed takes place by evaluation of

---

[1] Lewis, *An Experiment in Criticism*, 138.

its epistemic virtues, determined against a grid of independently established objective criteria. Yet a more critical reading of the history of science suggests that theories which are regarded as successful tend to act as validators for the criteria that were used in validating them in the first place. The effectiveness of a criterion of theoretical success is not determined on a priori epistemological grounds, but on its *a posteriori* success in validating a theory that is now believed to be successful.

There are, of course, difficulties with this approach. Hilary Kornblith is one of many philosophers of science to argue that there are no '*a priori* standards' of relevance to proper epistemic practice in the natural sciences.[2] The identification of 'appropriate inferential patterns' is an 'empirical affair'; the *legitimacy* of any scientific inference is dependent upon its *reliability*. Kornblith's cogent analysis helps us understand why so many philosophical considerations of the criteria of theory choice tend to focus on identifying 'reliable theories', and the criteria that were used to justify them.[3] The difficulty is, of course, that a theory that is judged reliable today may be discarded in the future, thus rendering this criterion historically contingent.

The study of scientific practice indicates that such archetypal 'rational' practices as scientific research and theorizing change over time, and that—despite popularizing assertions to the contrary—there are no 'generally applicable standards of rational acceptability in science'.[4] Instead, we find a 'roughly shared understanding of what can be assumed', reflecting the judgements of an epistemic community (or set of communities) concerning what is credible and reliable in the context of their ongoing work, in the light of the tasks to be undertaken and the resources at their disposal.

---

[2] Kornblith, *Knowledge and Its Place in Nature*, 21–3.

[3] For example, Plutynski, 'Parsimony and the Fisher-Wright Debate'; Baker, 'Occam's Razor in Science'.

[4] Rouse, *Knowledge and Power*, 124. For Rouse's rigorous defence of the centrality of scientific practice in shaping what is scientifically 'rational', see Rouse, *Engaging Science*, 125–57; idem, *How Scientific Practices Matter*, 263–58; idem, *Articulating the World*, 201–47. For criticism of Rouse's earlier work, see van Huyssteen, *The Shaping of Rationality*, 43–58.

WHAT IS A THEORY?

A theory is fundamentally 'a coherent description, explanation and representation of observed or experienced phenomena'.[5] The term is widely used in philosophy (e.g. 'theories of perception') and the humanities in general (e.g. 'Renaissance theories of translation'). In the natural sciences, the word is often used to designate such a way of understanding the world that has secured traction within the scientific community, although it is also used to refer to the views of individual scientists ('Newton's theory of planetary motion' or 'Dalton's atomic theory'), or to views that were once regarded as plausible within that community, but have since been discarded as a result of theoretical advance or the discovery of new evidence (such as 'phlogiston theory').[6]

Jurisprudence uses the terms 'legal doctrine' and 'legal theory', tending to see the former as the outcome of the application of the latter.[7] This usage is also often found in Christian theology, which tends to use the term *doctrina* to designate such a way of seeing the world as has found acceptance within a Christian community, while using 'theory' to designate a particular way of interpreting or applying this doctrine, alone or in combination with others, to gain a deeper understanding of the Christian vision of God.[8] Interestingly, nineteenth-century scientific works tend to use the word 'doctrine' as interchangeable with 'theory'—for example, referring to 'Dalton's atomic doctrine'. Charles Darwin thus often used the term 'doctrine' rather than 'theory', sometimes speaking of his 'doctrine of natural selection', where his modern interpreters would more naturally speak of a 'theory of natural selection'.[9] In part, this shift away from doctrine to theory represents a significant refocussing on how scientific ideas

---

[5] Lynham, 'Theory Building in the Human Resource Development Profession', 162.
[6] Woodcock, 'Phlogiston Theory and Chemical Revolutions'.
[7] Peczenik, 'A Theory of Legal Doctrine'.
[8] On which see van den Torren, 'Distinguishing Doctrine and Theological Theory'.
[9] See, for example, Darwin, *Life and Letters*, vol. 2, 155. Darwin rarely used the word 'evolution', preferring instead to speak of 'descent with modification'.

are conceived, in that the emphasis shifts from *understanding* to *perception*. A theory is about intellectual visualization of the world.[10]

A theory is a cognitive or imaginative framework or template, weaving together a series of known truths or proposed ideas into a coherent way of viewing the world, which is subsequently to be tested in terms of its ability to accommodate known and predict unknown observations. In some scientific situations, prediction is impossible—as, for example, in the case of Darwin's theory of natural selection, which is best seen as an explanation of past historical contingencies.[11] The capacity of a theory to predict novel observations is widely seen as indicative of its truth within the scientific community. Einstein's general theory of relativity, for example, posited a 'gravitational lens' created by the bending of light through warped space-time, which was observed during the solar eclipse of 29 May 1919. Other theories—such as string theory—have secured considerable traction within the scientific community on account of their mathematical elegance, yet in the conspicuous absence of experimental support.[12]

Theories arise through reflection on observations. This raises the question of the observations on which theology is grounded. One of the reasons for focussing the present discussion of rationality on Christian theology, rather than 'religion' in general, relates to the radically divergent understandings of the sources and norms of reasoning characteristic of individual religious traditions, and further internal divergence on these matters within each tradition. There are, for example, problems in correlating a philosophical monotheism (which argues for the existence of one single necessary being) and 'ancient Jewish monotheism', understood as the affirmation that there is only one deity who is properly to be worshipped.[13] The existence of other deities may be conceded, but the propriety of giving them worship is denied.

---

[10] For the background to this idea, see the perceptive study of MacKisack et al., 'On Picturing a Candle'.

[11] Hitchcock and Sober, 'Prediction vs. Accommodation and the Risk of Overfitting'. For the general issue, see Maher, 'Prediction, Accommodation, and the Logic of Discovery'.

[12] Penrose, *Fashion, Faith, and Fantasy in the New Physics of the Universe*, 1–10.

[13] Hurtado, *One God, One Lord*, 17–40; Bauckham, *Jesus and the God of Israel*, 107–25.

While Christianity can be correlated with classic and contemporary generic philosophical discussions about God—such as those associated with Plato and Aristotle—it nevertheless considers itself to be grounded in the religion of Israel, refocussed around the figure of Jesus of Nazareth.[14] Christianity is thus an interpretation of the identity and significance of Jesus of Nazareth, set out in the New Testament, and embodied, enacted, and transmitted through the community of faith, and expressed at different levels in its Creeds and public worship. There is a sense, therefore, in which theology can be seen as both the attempt to achieve the best intellectual articulation of the Christian faith, and the exploration of how such a Christian conceptual framework might enable both the meaningful inhabitation of our world, and the interpretation of our observations and experience. 'Christian doctrines arise out of attempts to understand and to do justice to our experience of Christ and of the Church, and not as airy items of unconstrained or ungrounded speculation.'[15]

The Greek term *theoria* affirms the importance of the manner in which we behold the world. We do not simply see the world, in whole or part; we see it *in a certain manner*, which is open to redirection and recalibration by a process of training and education.[16] A new theory thus allows us to see things in a new way. Thomas Kuhn makes this point in discussing the transition from Ptolemaic to Copernican theories of the solar system: 'Before it occurred, the sun and moon were planets, the earth was not. After it, the earth was a planet, like Mars and Jupiter; the sun was a star; and the moon was a new sort of body, a satellite.'[17] The phenomena were unchanged; they were, however, seen in a new light. Observation is thus not a neutral process, but is shaped by assumptions, explicit or implicit, about

---

[14] For an assessment of this process, see Hurtado, *Ancient Jewish Monotheism and Early Christian Jesus-Devotion*; Wright, *Paul and the Faithfulness of God*, vol. 2, 619–1042.

[15] Polkinghorne, 'Physics and Metaphysics in a Trinitarian Perspective', 37. For Polkinghorne's account of how Christianity is grounded in observation, see Polkinghorne, *Science and Christian Belief*.

[16] Adam, *Theoriebeladenheit und Objektivität*, 51–97.

[17] Kuhn, *The Road since Structure*, 15. For the impact of theory on categorization, see Gattei, *Thomas Kuhn's 'Linguistic Turn' and the Legacy of Logical Empiricism*, 144–6.

what is being observed.[18] One observer might see the sun rise and set; another might see the earth turning on its axis, leading to the apparent motion of the sun across the heavens.[19]

So how does one decide which is the best theory? Is there a 'theory of theory choice', so to speak, which helps us understand how individuals or communities gravitate towards a particular theory, or which articulates normative judgements about what criteria ought to be used in making and justifying such choices?[20] Are there verifiable criteria, based on empirical research, that should be deployed in this manner?[21] Or are these norms essentially pragmatic matters of judgement, determined by the values and working assumptions of a community of practitioners? In practice, it seems clear that there is no predetermined template of criteria; these rather emerge in the course of practice, and are judged largely in terms of how well they retrospectively evaluate theories that are known or believed to be successful, and might hence lead to the prediction of future successful theories. The history of science thus helps identify criteria that were used in the past by scientists to make theory choices which are now judged to have been correct, so that there are inductive grounds for operating within the constraints of these particular inductive criteria.[22] Thus the well-known philosophical difficulties with the inductive method are countered by the observation that, despite these problems, this method seems capable of leading to successful theoretical outcomes.

The recognition of the importance of epistemic values in theory selection dates from the late 1950s, as it became increasingly clear that scientists needed additional guidance for theory choice beyond simple criteria based on logic and evidence.[23] A careful reassessment of

---

[18] Adam, *Theoriebeladenheit und Objektivität*, 51–97.

[19] Radder, *The World Observed, the World Conceived*, 19–32.

[20] Thagard, 'The Best Explanation'. For the importance of this question to Thomas Kuhn's historicization of scientific rationality, see Kuhn, 'Objectivity, Value Judgment, and Theory Choice'; idem, 'Rationality and Theory Choice'.

[21] Achourioti, Fugard, and Stenning, 'The Empirical Study of Norms Is Just What We Are Missing'.

[22] This point is emphasized by Newton-Smith, *The Rationality of Science*, 224–5.

[23] See, for example, the influential studies of Churchman and Levi: Churchman, 'Science and Decision-Making'; Levi, 'On the Seriousness of Mistakes'. See also Kuhn, 'Objectivity, Value, and Theory Choice'.

earlier episodes in scientific history—such as Darwin's discovery of the principle of natural selection—which were often interpreted at the time in terms of logic and evidence has shown how they often anticipate later reflections on the role and reliability of certain rational virtues.

By the early 1980s, the phrase 'epistemic values' was beginning to gain acceptance as a means of designating the values that were regarded as acceptable in science as criteria for theory choice.[24] So what virtues might be seen as normative, or at least desirable, in guiding researchers to the most reliable interpretations of our world? Many such virtues have been proposed, in effect leading some to speak of the emergence of a repository of multiple epistemic virtues, with unresolved questions remaining concerning the manner and priority of their mutual relationships. The question of what might be judged to be the 'best' explanation of a set of observations will clearly depend on the nature of these criteria, and the manner in which they are applied.

We shall therefore turn to consider the general question of theory choice, focussing on the approach that has come to be known as Inference to the Best Explanation.

INFERENCE TO THE BEST EXPLANATION

The approach now generally known as 'Inference to the Best Explanation'[25] recognizes that multiple explanations might be offered for a set of observations, and sets out to identify criteria by which the best such explanation might be identified and justified.[26] It is inevitable that there will be multiple explanatory possibilities for any series of

---

[24] McMullin, 'Values in Science'.
[25] For an early statement of this approach, see Harman, 'The Inference to the Best Explanation'. The best study at present remains Lipton, *Inference to the Best Explanation*.
[26] Although 'inference to the best explanation' is sometimes confused with Peirce's concept of abduction, they should be seen as conceptually divergent: Minnameier, 'Peirce-Suit of Truth'; Campos, 'On the Distinction Between Peirce's Abduction and Lipton's Inference to the Best Explanation'.

observations, given the radical under-determination of theory by evidence.[27] It is possible to interpret 'Inference to the Best Explanation' as an extension of the notion of 'self-evidencing' explanations, in which the phenomenon that is to be explained itself provides reason for believing the explanation is correct. This situation frequently arises in the natural sciences, in that hypotheses are often supported by the observations that they are supposed to explain.

An example of this apparent circularity can be seen in a star's speed of recession explaining why its characteristic spectrum is red-shifted by a specified amount, despite the fact that this observed red-shift may actually be an essential element of the grounds which the astronomer has for believing that the star is receding at that speed.[28] Self-evidencing explanations exhibit a curious circularity, but this circularity is regarded as being benign—virtuous rather than vicious. The recession is used to explain the red-shift and the red-shift is used to confirm the recession, yet the recession hypothesis may be both explanatory and well supported. Inference to the Best Explanation thus partially inverts what might be considered as a natural or 'common-sense' view of the relationship between inference and explanation. According to this natural view, *inference is prior to explanation*. A scientist must first decide which hypotheses to accept, and will then draw from this group of accepted hypotheses to explain a given observation. Inference to the Best Explanation, however, holds that it is by only by asking how well various hypotheses would explain the available observational evidence that a scientist can determine which of those hypotheses merit acceptance. In this limited sense, Inference to the Best Explanation thus holds that *explanation is prior to inference*. This approach is, in a sense, about reasoning backwards.

Inference to the Best Explanation is thus really inference to the 'best' of the known and available competing possible explanatory hypotheses. It will thus be clear that the procedure of Inference to the Best Explanation is not about determining which of a given series of hypotheses is *true*; the objective is rather to determine which of

---

[27] Bonk, *Underdetermination*, 141–75; Laudan and Leplin, 'Empirical Equivalence and Underdetermination'.
[28] Lipton, *Inference to the Best Explanation*, 24–7.

these hypotheses functions as the 'best' explanation of what is observed, when assessed against a set of criteria—such as simplicity, elegance, and predictability. Harman, however, considered that there were good reasons for believing that the best explanation was likely to be true:[29]

> In making this inference, one infers from the fact that a certain hypothesis would explain the evidence, to the truth of that hypothesis. In general, there will be several hypotheses which might explain the evidence, so one must be able to reject all such alternative hypotheses before one is warranted in making the inference. Thus one infers, from the premise that a given hypothesis would provide a better explanation for the evidence than would any other hypothesis, to the conclusion that the given hypothesis is true.

Yet there is no agreed ranking of these criteria, and no agreement concerning how they might be generally applied (although there is much scholarly interest in how they were applied in the past to develop 'successful' theories). Paul Thagard, for example, engages in some detail with some important case studies, such as Darwin's argument in his *Origin of Species*; the wave theory of light, as developed by Huygens in the seventeenth century, and subsequently by Young and Fresnel in the nineteenth century; Newton's explanation of the motion of planets and satellites; and Halley's Newtonian prediction of the return of a comet.[30] Yet these illustrate the utility of these criteria, without illuminating the manner of their interaction, including their taxonomic ranking. Harman himself noted this problem, although he had little to say about how it might be addressed or resolved.[31]

There is, of course, a problem about how one is to judge that one hypothesis is sufficiently better than another hypothesis. Presumably such a judgment will be based on considerations such as which hypothesis is simpler, which is more plausible, which explains more, which is less ad hoc, and so forth. I do not wish to deny that there is a problem about explaining the exact nature of these considerations; I will not, however, say anything more about this problem.

---

[29] Harman, 'The Inference to the Best Explanation', 89.
[30] Thagard, 'The Best Explanation'.
[31] Harman, 'The Inference to the Best Explanation', 89.

The scientific utility of this method of reasoning is well established; but what of its theological counterparts?³² C. S. Lewis may have described himself as an 'empirical theist' who came to faith through 'induction'. Yet he is not typical. Relatively few religious believers come to faith through inductive or abductive processes of reasoning; most tend to speak of the motivation for their belief in terms of a response to, or encounter with, a personal, transcendent reality resulting in a commitment to a life of prayer, worship, and self-transformation. And having arrived at faith, many believers then turn to consider how their faith makes sense of what they observe and experience—not because it will lead them to faith, but because they wish to confirm or explore the sense-making capacity of their existing faith. This leads them to assess how a Christian *theoria* is able to accommodate their observations and experience—for example, the success of the natural sciences—and compare this with a rival way of seeing things, such as naturalism.³³ In developing this account of Christian rationality, Philip Clayton and Steven Knapp make two important points.³⁴

1. The Christian breaks no epistemic obligations by believing some things that have not been confirmed through intersubjective testing—such as a source accessible only to the Christian community, including the Bible or the Christian tradition;
2. The epistemic criterion for rationality is that all beliefs should be open to criticism in principle, and that those which are judged to be inadequate should be rejected. It is a theme, it may be added, which is familiar to any reader of the New Testament: 'test everything; hold fast to what is good' (1 Thessalonians 5:21).

There is little doubt that this mode of reasoning has considerable theological potential.³⁵ We have already mentioned C. S. Lewis's

---

³² For excellent discussions, see Clayton, 'Inference to the Best Explanation'; Reichenbach, 'Explanation and the Cosmological Argument'; van Holten, 'Theism and Inference to the Best Explanation'.
³³ Clayton and Knapp, 'Rationality and Christian Self-Conception', 133–4. For an excellent account of belief as a 'disposition towards judgement', see Ward, *Unbelievable*.
³⁴ Clayton and Knapp, 'Rationality and Christian Self-Conception', 135.
³⁵ For an early indication of its potential, see Prevost, *Probability and Theistic Explanation*. A more explicitly theological approach is found in Clayton, 'Inference to the Best Explanation'.

appreciation of the accommodative capacities of the Christian *theoria*, which allows it to 'fit in' art, the natural sciences, morality, and other religions.[36] At a more explicitly philosophical level, William J. Wainwright follows Basil Mitchell in arguing that Christianity may be seen as a worldview or metaphysical system which attempts to make sense of human experience as a whole, and uses criteria similar to those used to judge other forms of explanation. For Wainwright, good metaphysical theories of this kind should meet three formal criteria. They must be logically consistent; they should be coherent, displaying a certain amount of internal interconnectedness and systematic articulation; and they should be simple, rather than complex.[37] Other criteria might be adduced for the adequacy of a theological system which do not relate to its explanatory functions—such as its faithfulness to authoritative texts such as the Christian Bible, its acceptance by the *consensus fidelium*, or its perceived existential adequacy.

There is a clear resonance with this approach in C. S. Lewis's formulation of an 'argument from desire'. What, he asks, is 'the most probable explanation' of 'a desire which no experience in this world can satisfy'?[38] Lewis sets out three possible explanations of our experience of desire, especially of our sense of emptiness and lack of fulfilment, and indicates which he considers to be the best. The first such explanation is that this frustration arises from looking for its true object in the wrong place; the second is that there is no true object of desire to be found. If this second explanation is true, then further searching will result only in repeated disappointment, suggesting that there is no point in trying to find anything better than or beyond the present world.

Lewis then suggests that there is a third approach, which recognizes that these earthly longings are 'only a kind of copy, or echo, or mirage' of our true homeland. Since this overwhelming desire cannot be fulfilled through anything in the present world, this suggests that

---

[36] Lewis, *Essay Collection*, 21. For Lewis, the 'scientific point of view cannot fit in any of these things'—including the success of science itself.
[37] Wainwright, 'Worldviews, Criteria and Epistemic Circularity'. For similar stipulations, see Brümmer, 'The Intersubjectivity of Criteria in Theology'.
[38] Lewis, *Mere Christianity*, 136-7.

its ultimate object lies beyond the present world. 'The most probable explanation is that I was made for another world.'[39]

Lewis makes it clear that this is not about 'proving' anything; it is about trying to identify which, of several possible explanations, is the best, or the 'most probable'. For Lewis, this third is the 'most probable' explanation (although he does not clarify the criteria by which this probabilistic judgement might be made). Lewis clearly sees his analysis of the human experience of desire as an expression of the rationality of faith. Conceding that other explanations of this experience are indeed possible, he argues that the Christian explanation is the best.

This approach can be extended to engage other observations. Consider, for example, the increasingly widespread recognition that it is natural to believe in God.[40] It remains unclear whether this constitutes evidence for or against religion, and it certainly does not constitute a proof of theism.[41] Yet Lewis's approach suggests that such a natural inclination can be accommodated convincingly within a theistic perspective—for example, though conceiving the 'image of God' in terms of a relational homing instinct for the transcendent and meaning,[42] captured in the descriptor *homo religiosus*.

CORRESPONDENCE AND COHERENCE AS THEORETICAL VIRTUES

Theories, whether scientific or theological, aspire to possess and exhibit both an extra-systemic and an intra-systemic rationality.[43] Scientific theories must be grounded in the real world, and aim at

---

[39] Lewis, *Mere Christianity*, 136-7.
[40] See, for example, Barrett, *Why Would Anyone Believe in God?*; idem, *Born Believers*.
[41] As noted in Jong, Kavanagh, and Visala, 'Born Idolaters'.
[42] Shults, *Reforming Theological Anthropology*, 117-39.
[43] For the general issue, see Hornbostel, *Wissenschaftsindikatoren*, 21-76. On some theological aspects of this question, see Trenery, *Alasdair Macintyre, George Lindbeck, and the Nature of Tradition*, 201-7.

maximal internal coherence.[44] They are accountable to the reality they purport to represent. 'Science cannot dispense with extra-systematic concepts: if there were no links between systematic and extra-systematic concepts, scientific theories would be untestable and unintelligible.'[45] Ontological finality is thus understood to rest with nature itself. This viewpoint is widespread within the natural sciences, which have generally disregarded anti-realist arguments on account of their failure to account adequately for the success of scientific theorization and explanation. Following the lines of the approach of Hilary Putnam, most scientists argue that the success of science would be a miracle if our theories were not at least (approximately) true.[46]

Yet scientific theories are not merely understood to be grounded in an external reality; the ideas which arise from an engagement with reality should ultimately be consistent with each other. A good theory is understood to 'correspond coherently to reality', in that it is grounded in an engagement with the real world, and is internally coherent.[47] Neither internal coherence nor some form of correspondence with reality is adequate in itself.

While this kind of approach finds widespread acceptance within the natural sciences, it also has significant theological traction. Wolfhart Pannenberg's early theological project took the form of demonstrating the internal coherence of Christian doctrines on the one hand, and the external coherence or consistency of those doctrines with the world of reality and other intellectual disciplines on the other.[48] Similarly, the Anglican theologian Charles Gore, while affirming the importance of the 'coherence of Christian doctrine', did not hold that the plausibility of Christian doctrine rested upon that coherence alone; it was to be judged in terms of its reliability in conveying the Christian understanding of the significance of Jesus

---

[44] Dawson and Gregory, 'Correspondence and Coherence in Science'.
[45] Bunge, *Philosophy of Science*, 100.
[46] For discussion, see Agazzi, *Scientific Objectivity and Its Contexts*, 243–312; Psillos, *Scientific Realism*, 72–93.
[47] Thagard, 'Coherence, Truth, and the Development of Scientific Knowledge'; Kitcher, 'On the Explanatory Role of Correspondence Truth'.
[48] Pannenberg, *Systematic Theology*, vol. 1, 21–2.

Christ, as presented in Scripture and the Christian experience.[49] Some theologians have attempted to disentangle extra-systemic correspondence and intra-systemic coherence, seeing doctrine primarily in terms of the internal regulation of Christian language.[50] Yet classical Christian theology has affirmed that doctrine is more than the internal regulation of a community's language of faith; it is about offering a theory—a way of seeing things—which represents both an internally coherent theoretical representation and a corresponding external reality.

It is possible to argue that, seen in terms of the historical process of its development, Christian doctrine was primarily concerned with articulating and safeguarding intra-systemic consistency. Athanasius of Alexandria, for example, argued that the internal intellectual coherence of Christianity was compromised by Arius's Christology, which was clearly inconsistent with a core element of Christian practice in the fourth century—the worship of Christ.[51] Yet while an outsider might see Christianity as grounded upon and characterized by its own controlling narrative, capable of generating a set of distinctive ideas, Christianity has seen this as, not simply a *narrative*, but a *metanarrative*.[52] Both J. R. R. Tolkien and C. S. Lewis regarded Christianity as a 'true myth', telling a story that places and accounts for other stories, and subsequently generating a set of ideas derived from this story.[53] There is no inconsistency here; any tradition-mediated rationality or metanarrative is called upon to account for the existence of rival accounts of rationality and alternative narrations

---

[49] Gore, *The Incarnation of the Son of God*, 21–6, 96–106.

[50] This criticism is often directed against George Lindbeck's intra-systemic approach to doctrine: see O'Neill, 'The Rule Theory of Doctrine and Propositional Truth'. Lindbeck, however, is not entirely consistent here, and suggests that his cultural-linguistic model of doctrine does not entail the rejection of either an epistemological realism or a correspondence theory of truth: Lindbeck, *The Nature of Doctrine*, 68–9.

[51] Williams, *Arius*, 95–115, especially 110. For the general issue, see Lehmkühler, *Kultus und Theologie*, 208–32.

[52] Watson, *Text, Church and World*, 82–4.

[53] McGrath, 'A Gleam of Divine Truth'. The ideas of the German philosopher Kurt Hübner are of interest here, not least in correlating the categories of 'myth' and 'rationality': see Rieger, 'Grenzen wissenschaftlicher Rationalität, Relativismus und Gottesglaube'.

of identity and meaning—in other words, to account for something that exists outside that community, in terms of that community's own framework of meaning.[54]

The roots of this approach are complex, and lie partly in the expansion of Christianity during the period of the early church. Much theological discussion initially focussed on the relation of Christianity to Judaism, attempting to identify the nature of the continuity and discontinuity between the two movements.[55] Yet as Christians became familiar with both Hellenized versions of Judaism (such as the writings of Philo) and Hellenistic philosophy, a growing interest developed in Plato's notion of *logos*, and the way in which this could be correlated with the Christian notion of Jesus Christ as the incarnate logos.[56] The notion of God creating a 'rational (*logikos*)' world created a growing interest in showing how Christianity could engage and even surpass existing philosophical discussions about the nature of the world and the divine. Without losing its intra-systemic focus, Christian theology developed an extra-systemic role, as the apologetic potential of the explanatory capacity of faith began to be appreciated.

This development was consolidated in the Latin-speaking West, in Augustine of Hippo's emphasis upon the human intellect's intrinsic ability to cognize in a unifying manner. For Augustine, faith opens, heals, or cleanses the eyes of the mind, allowing the believer to see things as they really are, and finally to behold God.[57] Augustine uses the model of illumination, not entirely consistently, to indicate the ways in which God makes it possible for human beings to see themselves and the world as God sees them. These ideas were developed further in the Western theological tradition, and helped shape its understandings of rationality.

This development had some important implications. In the first place, Christianity understands its 'big picture' as capable of engaging and making sense of the wider world, so that the Christian community

---

[54] For a detailed reflection on this general issue, see Detjen, *Geltungsbegründung traditionsabhängiger Weltdeutungen im Dilemma, Theologie, Philosophie, Wissenschaftstheorie und Konstruktivismus*.
[55] Lieu, *Christian Identity in the Jewish and Graeco-Roman World*, 98–146.
[56] Edwards, *Image, Word and God in the Early Christian Centuries*.
[57] For an excellent analysis, see Schumacher, *Divine Illumination*, 25–65.

is able to offer its own distinct account of reality. Christian theology is thus not a purely self-referential system; it has, in principle, a capacity to explain and account for what lies outside its own domain. It is therefore theologically legitimate and productive to explore how Christianity generates new ways of seeing our world and ourselves, and consider how this is reflected in the values Christians attach to them. It suggests that, as we shall discuss in more detail later (140–3), an epistemic mode of explanation is particularly adapted to Christian theology, in that this generates a map of meaning within which events or entities may be located.

Yet in the second place, this suggests that Christianity is under an intellectual obligation to defend its own distinct rationality, in terms of demonstrating that it can offer either the 'best' explanation of the world, or at least an adequate account. While Christian apologetics takes many forms,[58] one of its most characteristic approaches consists of the affirmation of its own distinct rationality, and showing how this allows both observation and experience to be interpreted in a meaningful and at least plausible manner. The granularity of a tradition's interpretation of the world is thus seen as an indicator of its reliability, as well as of its appeal beyond the boundaries of its own community.

OBJECTIVITY

In his magisterial project *Science and the Shaping of Modernity*, Stephen Gaukroger offers an account of the 'fundamental transformation of intellectual values' that are held to be constitutive of the modern era. One of the most important elements of this transformation of intellectual culture, especially in the natural sciences, is the emergence of the virtue of objectivity, which aims for an impartial and detached account of reality which could be accepted by any individual, in that it does not draw on any assumptions, prejudices,

---

[58] For a good account of the origins and form of this development in the early church, see Dulles, *A History of Apologetics*, 27–90; Pelikan, *Christianity and Classical Culture*, 3–151.

or values.⁵⁹ This virtue—evident in Francis Bacon's notion of scientific self-distancing,⁶⁰ but seemingly contradicted by Bacon's personal inclinations to adapt his views for financial considerations⁶¹—has a history, in that it does not possess the status of a 'universal given', but has rather been constructed and altered over the centuries, especially during the transition to modernity.⁶² Recognition of this has increasingly led intellectual historians to treat 'objectivity' and cognate notions as 'historical products' or 'actors' judgments', rather than self-evidently appropriate values for, or definitional prerequisites of, the scientific enterprise.⁶³

Yet these observations on the historical emergence of the epistemic virtue of objectivity do not negate its utility. It is not difficult to understand the appeal of disinterested and impartial observers of our world,⁶⁴ who are able to don a 'veil of ignorance' that prevents them from knowing their race, social status, gender, religion, talents, and other defining characteristics or vested interests that might skew their judgements about the morality or wider implications of any decision. Such an aperspectival objectivity clearly has its virtues; nevertheless, it runs the risk of being a 'view from nowhere'. in effect being an idealistic construction rather than a practical reality. Perhaps for this reason, many have argued that objectivity is best achieved communally, rather than individually. Communal approaches to truth, it could be argued, proceed by a cancellation of individual biases or an elimination of idiosyncrasies and the distorting influences of personal histories.

Yet speaking of 'objectivity' might suggest that observation is invariably or necessarily a cool and detached process, in which the observer is disengaged from what is being observed. The beholding of certain objects can arouse, excite, disturb, and inspire the observer. In his *Modern Painters*, John Ruskin famously contrasted what he termed *aesthesis* with the more committed and informed activity of

---

⁵⁹ Gaukroger, *The Emergence of a Scientific Culture*, passim.
⁶⁰ Zagorin, 'Francis Bacon's Objectivity and the Idols of the Mind'.
⁶¹ Stewart, 'Bribery, Buggery, and the Fall of Lord Chancellor Bacon'.
⁶² Daston and Galison, *Objectivity*, 27–35.
⁶³ Solomon, *Objectivity in the Making*, 1–8. An early example of this approach is Shapin and Schaffer, *Leviathan and the Air-Pump*, 13–14.
⁶⁴ Adam, *Theoriebeladenheit und Objektivität*, 179–244.

theoria. For Ruskin, *aesthesis* was the 'mere animal consciousness' of beauty, whereas *theoria* was the 'exulting, reverent and grateful perception' of that same beauty.[65] At times, Ruskin conceived *theoria* in almost religious terms, as a sacred beholding of the transcendent, perhaps hinting at the theological associations of the term in ancient Greece.[66] Ruskin offers an important corrective to Martin Heidegger's suggestion that *theoria* entails the notion of 'spectatorial vision' which leads to the distancing of the knowing subject from the known object, thus leading to their mutual estrangement.[67] Although Ruskin had yet to clarify his understanding of the relation of the knowing subject and the known object, especially between the externally given and the internally modified imaginative representation, he saw no reason to divorce them.

This leads into a potentially significant divergence between scientific and religious reflections on theory. While there is some danger in this generalization, scientific theorizing is primarily concerned with achieving an enhanced understanding of the natural world; theological theorizing, while also aiming for a deeper understanding of God and the world, is seen to lead seamlessly into the praxis of adoration and prayer. There is an intimate connection between theology and worship (91–2), involving the coordination of objective and subjective elements, going beyond the traditional acknowledgement that the practice of worship informs the content of theology (*lex orandi, lex credendi*).[68] Theology articulates a vision of God which cannot be adequately accommodated by the human intellect, and thus generates a sense of intellectual wonder most appropriately expressed in worship. We shall return to this point in a later chapter, when we consider the concept of mystery in science and theology.

Yet many scientists, whether religious or not, would suggest that the conventional account of science as an attempt to understand or

---

[65] Ruskin, *Works*, vol. 4, 42. Ruskin was not entirely consistent in his exposition of this notion of *theoria*. For a more recent exposition and application of this idea, see Fuller, *Theoria: Art and the Absence of Grace*.

[66] Rausch, *Theoria: Von ihrer sakralen zur philosophischen Bedeutung*, 148–52.

[67] McGrath, *Re-Imagining Nature*, 58–9.

[68] Clerck, 'Lex Orandi, Lex Credendi'; Schneider, *Zur theologischen Grundlegung des christlichen Gottesdienstes nach Joseph Ratzinger*.

explain our world is deficient, not simply in terms of its resulting attitude to that world, but because it fails to do justice to the actual attitudes of scientists themselves. An experience of a sense of awe in the presence of nature transcends any attempt to reduce it to verbal or conceptual formulae. Such an experience of awe, although 'often fleeting and hard to describe', diminishes the emphasis on the individual self and enables a more expansive vision of our world, which cannot be reduced to words or formulae, and which opens individuals to new modes of thought and action.[69]

In affirming, although critically, the importance of objectivity as a virtue in the scientific method, it is important to note the challenge it poses to what might be termed a 'feels-right' diagnostic standard for the acceptance of an explanation.[70] Yet there is a subjective element in scientific judgement, as the widespread appeal to beauty or elegance in theory choice indicates. We shall consider this virtue later in our discussion (117–19).

SIMPLICITY

The human cognitive system is required to cope with a world that is at one and the same time immensely complex yet highly patterned. In a completely random world, there would be no patterns to be observed, so that making the scientific processes of prediction, explanation, and understanding would be impossible. Human cognition thus involves the discernment of patterns of association or behaviour which can serve in explanatory or predictive manners. Some psychologists have argued that the human cognitive system imposes patterns on the world according to a 'simplicity principle', in which the chosen pattern provides the most compact representation of the available information.[71] In practice, the application of this 'simplicity principle' is normatively justified in that it generally leads to simple

---

[69] Shiota, Keltner, and Mossman, 'The Nature of Awe'; Piff et al., 'Awe, the Small Self, and Prosocial Behavior'.
[70] Trout, 'Scientific Explanation and the Sense of Understanding', 214.
[71] See especially Chater, 'The Search for Simplicity'.

representations that provide good explanations and predictions on the basis of which the agent can make decisions and actions. Willard van Quine seems to have anticipated such considerations when he suggests that our perceptions of what is 'simple' may in fact arise from contingent features of our perceptual mechanisms.[72]

The assertion that simplicity is a *sign*—but most emphatically not a *guarantor*—of truth has a long history of use in the natural sciences. In inscribing the Latin motto *simplex est sigillum veritatis* on the flyleaf of his *Entwickelungsgeschichte der Thiere* (1828), the pioneer embryologist Karl Ernst von Baer was perhaps doing more than echoing the widespread intuition of a direct correlation between simplicity and truth; he was also criticizing the biogenetic doctrine of recapitulation, which held that the development of the embryo of a higher animal passes through stages which resemble or recapitulate successive stages in the evolution of the animal's distant ancestors. The principle of parsimony is regularly referenced in evaluating the merits of competing explanations (as in contemporary debates over the evolutionary theories of R. A. Fisher and Sewall Wright.)[73] So is simplicity a reliable guide to the validity of a theory?

Karl Popper certainly thought so, declaring that 'simple statements' are to be 'prized more highly than less simple ones *because they tell us more; because their empirical content is greater; and because they are better testable*'.[74] Popper's surprisingly positive estimate of the virtue of simplicity, of course, rests chiefly on his controversial equation of this concept with the degree of falsifiability of a theory. Yet it is fair to suggest that the history of the natural sciences lends some support to the general principle of avoiding conceptually inflated theories. William of Ockham's advice—summed up in maxims such as *numquam ponenda est pluralitas sine necessitate* and *frustra fit per plura quod potest fieri per pauciora*—to avoid the unnecessary multiplication of hypothesis remains valuable; it is not, and was not intended to be, a touchstone of truth. Modern formulations of 'Ockham's Razor' often have little conceptual connection with Ockham's own approach,

---

[72] Quine, 'On Simple Theories of a Complex World'.
[73] Plutynski, 'Parsimony and the Fisher-Wright Debate'. Other applications may be noted, such as Baker, 'Occam's Razor in Science'.
[74] Popper, *The Logic of Scientific Discovery*, 128. Emphasis in original.

which tended to be directed against such conceptually inflationary notions as a created habit of grace.[75] Isaac Newton's first 'rule of reasoning in philosophy' was 'to admit no more causes of natural things than such as are both true and sufficient to explain their appearances'.[76] This approach allows a theory to be envisaged as a pattern, a way of encoding observational data, which leads to the suggestion that the pattern chosen should be that which allows the simplest encoding of this data.[77] Yet the progress of scientific knowledge does not invariably lead to the simplest theories. For example, Kepler showed that the planets did not orbit the sun in mathematically simple circles, but in the more complex curves of ellipses, requiring more complex mathematical representation. Simplicity cannot be upheld as a general indicator of theoretical validity, any more than complexity can be regarded as a criterion of falsity.[78]

Richard Swinburne, who makes extensive use of an appeal to simplicity in support of theistic arguments, asserts that 'it is an ultimate *a priori* epistemic principle that simplicity is evidence of truth'.[79] It is far from clear that this is either an a priori principle, nor that it can be seen as 'evidence of truth'. A more accurate account of the situation would be that the comparative success of using simplicity as a criterion of scientific theories is an a posteriori indication that it is an indicator of truth. It is difficult to see how simplicity can be advocated as an a priori criterion of truth without a set of informing metaphysical assumptions, which themselves add metaphysical complexity to such a theoretical evaluation.

While many believe that the simplest theory is likely to be the best, the ultimate foundation of this belief thus remains unclear.[80] Nor is it clear whether this privileging of simplicity is really an aesthetic intuition, rather than a validated empirical tool. Is 'simplicity' being

---

[75] For the medieval debates, see McGrath, *Iustitia Dei*, 176–86.
[76] Christianson, *Isaac Newton and the Scientific Revolution*, 86.
[77] Chater, 'The Search for Simplicity', 278.
[78] Bunge, 'The Weight of Simplicity in the Construction and Assaying of Scientific Theories'; Elkana, 'The Myth of Simplicity'.
[79] Swinburne, *Simplicity as Evidence for Truth*, 1.
[80] For example, it could be argued that simplicity enhances truth-finding efficiency: Kelly, 'A New Solution to the Puzzle of Simplicity'. For criticism, see Fitzpatrick, 'Kelly on Ockham's Razor and Truth-Finding Efficiency'.

understood as a *logico-empirical* criterion, or as an essentially *aesthetic* judgement?[81] And how is the simplicity of a theory to be measured, so that it becomes meaningful to speak of one theoretical analysis as being 'simpler' than another?[82] Although the general human preference for simple patterns has been widely recognized, simplicity has nevertheless remained a largely intuitive notion,[83] perhaps even a matter of personal taste. The deeper question here is whether a theoretical adjudication is being made on the basis of a subjective value or a measurable quality—an issue to which we shall return in considering the role of beauty as a criterion of theory choice (117–19).

So how might this criterion of simplicity bear upon discussions of rationality in science and theology? Eliott Sober suggests that three issues need to be resolved before this criterion can be applied meaningfully: how simplicity is to be measured; how it can be justified; and how it should be 'traded-off'—that is, how it is to be weighed against other theoretical virtues.[84] None of these are easily answered. For example, many argue, in a common-sense manner, that a simpler theory is more likely to be true. Yet this is essentially a pragmatic judgement, not an argument for the epistemic grounds of simplicity.

As noted earlier, Swinburne argues that 'simplicity is evidence for truth' in his assertion of the plausibility of the existence of God as a sense-making device. This might be argued to constitute proper grounds for the rationality of faith; it is, however, more satisfying and realistic to see this as consistent with the existence of God, or congruent with a theistic logic, particularly the distinctive Trinitarian logic of Christianity, which holds that God created an ordered and elegant world. Others, however, have argued that the criterion of simplicity is intrinsically anti-theist or non-theist, in that postulating the existence of God requires the addition of one item to the existing

---

[81] Walsh, 'Occam's Razor: A Principle of Intellectual Elegance'.
[82] Hillman, 'The Measurement of Simplicity', McAllister, 'The Simplicity of Theories'.
[83] The concept of 'Kolmogorov Complexity' clearly has potential in this respect, although this has been under-developed to date. See Grünwald and Vitányi, 'Kolmogorov Complexity and Information Theory'.
[84] Sober, 'What Is the Problem of Simplicity?'

inventory of the universe, hence resulting in a more complex theory.[85] Theists respond by pointing out that their notion of God is inherently explanatory; this generative concept of God thus gives rise to, if anything, a *reduction* in the number of explanatory elements required to make sense of the universe.[86]

Yet problems with the use of the criterion of simplicity remain. It is difficult to define and operationalize the notion, and to provide it with an independent epistemic foundation. Furthermore, it is by no means clear that simplicity is a sign of truth, or even an indicator of the potential long-term success of a theory. Perhaps unsurprisingly, many philosophers of science now tend to see simplicity therefore as a desirable quality for theories, while recognizing that many theories deemed to be valid or successful are not simple.

ELEGANCE AND BEAUTY

Logico-empirical approaches to theory choice extol the importance of certain virtues, including consistency with known observations, a capacity to predict, explanatory power, and internal consistency. Yet many leading theoreticians have emphasized that beauty or perceived elegance appears to serve as an indicator of theoretical truth. 'The affirmation of a great scientific theory is in part an expression of delight. The theory has an inarticulate component acclaiming its beauty, and this is essential to the belief that it is true.'[87] The theoretical representations of the natural world often impress observers as beautiful in themselves. The conceptual elegance of Euclidean geometry was long regarded as one of its chief virtues. More recently, Newton's dynamic equations were widely admitted for their conceptual elegance, displayed even more strikingly by the symplectic geometry of the Hamiltonian or Lagrangian formalisms.[88]

---

[85] This appears to be the argument in Dawkins, *The God Delusion*.

[86] Inge, *Faith and Its Psychology*, 197. For Kepler's appeal to theological notions in articulating the unity of nature, see Hon, 'Kepler's Revolutionary Astronomy', 162–73.

[87] Polanyi, *Personal Knowledge*, 133. For the general issues, see McAllister, 'Truth and Beauty in Scientific Reason'; Barnes, 'Inference to the Loveliest Explanation'; Kivy, 'Science and Aesthetic Appreciation'.

[88] Penrose, *The Road to Reality*, 483–91.

If beauty does indeed occasionally seem to function as a criterion of theory choice, the theoretical basis of this function must be said to remain elusive. It might be argued that scientists have come to associate 'beauty' with 'truth' on account of previous experience of theoretical situations in which this connection proved fruitful.[89] The association in question is therefore not arbitrary, in that it is grounded in the features of past successful theories, and the practices of scientific communities. Or perhaps the perception that a theory is elegant or beautiful actually reflects its perceived coherence,[90] so that beauty is a place-holder for some other quality more easily harmonized with an essentially logico-empirical approach to theory choice.

Yet the concept of 'beauty' is subjective and contested, leading some to make the 'eminently rational decision' to pursue 'indicators of truth in disregard of beauty'.[91] Properties of a theory that have at some point been considered to be aesthetically attractive have at other times been considered neutral or displeasing. This suggests to some that scientific theories gain acceptance on the basis of essentially rational criteria, with aesthetic criteria being involved subsequently, often as retrodictive explanations of the theory's success. The correlation between aesthetic and epistemic values remains unclear. It is an empirical truth that there sometimes appears to be some correlation between the perceived beauty of a theory or theoretical representation of the natural world and its veridicality; yet the nature and extent of this correlation are sufficiently unclear and unreliable to warrant scepticism about its capacity to adjudicate between theoretical possibilities. Any connection between 'beauty' and 'truth' has to be provided by an additional informing theory, implicit or explicit, in that such a connection is not itself to be seen as a secure empirical correlation.

In practice, most of those who equate or elide 'beauty' and 'truth'—or allow them to be seen as functionally equivalent—do so on the basis of an unstated or unacknowledged set of assumptions about the

---

[89] Kuipers, 'Beauty, a Road to the Truth'.
[90] Thagard, 'Why is Beauty a Road to Truth?'
[91] McAllister, 'Truth and Beauty in Scientific Reason', 45.

nature of beauty, which are ultimately not empirical. 'Beauty serves us as a schema for truth, a postulated substitute for a reality which we cannot fathom.'[92] For many physicists, beauty appears to be framed primarily in terms of symmetry,[93] as exemplified in the truncated icosahedrons of certain classes of allotropes of carbon. Now symmetry may indeed be one element of any theoretical account of beauty; it cannot, however, be considered to be determinative of the notion, not least because of its implicit reduction of beauty to a 'sterile rigidity'.[94]

Some writers appeal to explicitly theological notions in their correlation of truth and beauty, a theme which played an important role in the Scottish geologist Hugh Millar's reflections on the deeper significance of the beauty of natural forms.[95] Yet perhaps the more important function of a theological perspective is that viewing the world through a Trinitarian *theoria* allows us to evade the criticism directed by Erich Heller against Nietzsche's sense of awe in the presence of nature—namely, that it amounted to a *religio intransitiva*, based on a sense of awe which had no awesome referent.[96]

A CAPACITY TO PREDICT

A sharp distinction is often drawn between the capacity of scientific theories to predict novel observations, and the purely retrodictive capacities of religious theories. A capacity to accommodate observation and experience is, of course, an integral element of a successful theory, whether scientific or religious. Yet a theory which proves capable of predicting something hitherto unobserved is widely regarded as trumping those which merely account for what is already known.[97] If E designates evidence which might be held to support

---

[92] Zemach, *Real Beauty*, 36.
[93] A classic statement of this is found in Weyl, *Symmetry*.
[94] McManus, 'Symmetry and Asymmetry in Aesthetics and the Arts'.
[95] Brooke, 'Like Minds: The God of Hugh Miller'.
[96] Heller, *The Disinherited Mind*, 171.
[97] For a good account of the debate, see White, 'The Epistemic Advantage of Prediction over Accommodation'.

theory T, then it is often asserted that E offers greater support for T if it is a novel prediction than a mere accommodation. The capacity to predict is thus a comparative tool, based on a judgement of the relative probative weight of E as predicted when compared to the probative weight of E as accommodated.[98]

It seems entirely plausible that such a generative theory is to be preferred over one that merely casts a theoretical net over what is already known. Yet it is important to appreciate that this epistemic virtue has a social history, being particularly linked with August Fresnel's defence of a wave theory of light, challenging the prevailing Newtonian corpuscular theory. Siméon Denis Poisson pointed out that, if Fresnel's theory was right, it followed that if a small circular disk was used to create a shadow caused by a point source of light, the centre of that shadow would contain a white spot of light.[99] This totally counter-intuitive prediction, which Fresnel himself had failed to notice, was seen by Poisson as an indication that Fresnel's theory was wrong. François Arago, a colleague of Poisson, immediately set out to investigate this predicted phenomenon—and observed it. The high drama of this event served to consolidate the view that the capacity of a theory to predict would, when the predictions were observed, represent a startlingly persuasive demonstration of its validity.

Yet it proves difficult to identify precisely why a capacity to predict should be seen as epistemically virtuous.[100] The critical question is whether evidence is supportive of a theory, and the rigour of the selection procedure used to generate the evidence. The chronological ordering of theory and evidence does not ultimately affect the veracity of a theory. This issue emerged as important in the nineteenth-century debate between William Whewell and John Stuart Mill over the role of induction as a scientific method.[101] Whewell emphasized the importance of predictive novelty as a core element of the scientific method; Mill argued that the difference between prediction of novel observations and theoretical accommodation of existing observations was purely psychological, and had no ultimate epistemological significance.

---

[98] Barnes, *The Paradox of Predictivism*, 1–7.
[99] See especially Worrall, 'Fresnel, Poisson and the White Spot'.
[100] Achinstein, *The Book of Evidence*, 210–30; Gonzalez, *La predicción científica*.
[101] Snyder, 'The Mill-Whewell Debate'.

Rational Virtues and the Problem of Theory Choice 121

Furthermore, the nature of certain scientific fields of explanation is such that they cannot predict in any meaningful sense. Recent studies in the philosophy of biology have raised interesting questions about whether prediction really is essential to the scientific method. While prediction is superior to accommodation in many cases (particularly when such accommodation is seen to be 'fudged', contrived, or forced), this is not always so.[102] And certain areas of science are resistant to predictivist approaches, on account of the nature of the phenomena being observed.

Prediction is not *essential* to the scientific method. The archaeologist who studies Neanderthal sites may gain an understanding of the early genetic history of human beings, which may be enlightening, but is not predictive in any meaningful sense of the term. A fluid dynamicist studying the chaotic outcome of convection processes in a Bénard cell may well understand the basic principles underlying these processes, but knows that the specific forms that such chaotic phenomena take are, by their very nature, fundamentally unpredictable.[103] Understanding general scientific principles or universal laws does not generally lead to a deterministic prediction of behaviour.

Charles Darwin was quite clear that his theory of natural selection did not predict, and could not predict. It could not be proved, but was to be judged by its capacity to accommodate the evidence.[104] In a letter praising the perspicuity of F. W. Hutton (1836–1905), Darwin singled him out for special comment, in that 'he is one of the very few who see that the change of species cannot be directly proved, and that the doctrine must sink or swim according as it groups and explains phenomena'.[105] This absence of predictive capacity, of course, led

---

[102] Hitchcock and Sober, 'Prediction versus Accommodation and the Risk of Overfitting'. The 'weak predictivism' defended by Hitchcock and Sober has parallels elsewhere: see, for example, the careful assessment of approaches in Lange, 'The Apparent Superiority of Prediction to Accommodation as a Side Effect'; Harker, 'Accommodation and Prediction'.
[103] For the importance of random events for predictive capacity, see Eagle, 'Randomness Is Unpredictability'.
[104] See especially the detailed study by Lloyd, 'The Nature of Darwin's Support for the Theory of Natural Selection'.
[105] *The Life and Letters of Charles Darwin*, vol. 2, 155. Hutton deserves much greater attention as a perceptive interpreter of Darwin: see Stenhouse, 'Darwin's Captain'.

some philosophers of science, most notably Karl Popper, to suggest that Darwinism was not really scientific.[106]

Perhaps, as some suggest, Darwin's theory of natural selection may primarily have given 'structure to our ignorance';[107] yet it also postulated a plausible mechanism for evolution which could give rise to a fruitful research programme. The nature of those biological observations was such that a capacity for prediction was not a possible criterion of evaluation for Darwin in relation to his theory of natural selection. The capacity of a theory to predict is shaped to no small extent by the specific object of its investigation. In some cases, prediction is possible and virtuous; in others, it is not. This does not reflect any failures of the rationality of the process of theory choice; it is simply a recognition of the complexity of the realities that theories are called upon to explain and connect.

Religious theories do not predict in the manner of some scientific theories;[108] this, however, is not to be seen as a failure in rational virtue on their part, but as a specific outcome of the object of theological study, and a reflection of the limiting conditions under which theology is obliged to proceed.

Thus far, we have considered a number of criteria that might help assess the reliability of a proposed theory. Others, of course, could be added, including a range of different epistemic virtues and vices—such as curiosity and humility, arrogance and dogmatism—and exploring how they affect a person or group's capacity to acquire, assess, and apply knowledge through standard epistemic practices, such as developing programmes of research, theorizing about their outcomes, and debating with those who hold rival positions.[109] The emerging field of 'virtue epistemology' highlights the importance of the character of an investigation, noting the importance of certain virtues (such as humility, openness, and a willingness to be guided by evidence) and their corresponding vices (such as arrogance, dogmatism, and closed-mindedness) in guiding the process of enquiry.[110]

---

[106] Popper, 'Natural Selection and the Emergence of Mind'. Popper later retracted this criticism.
[107] The view of Kitcher, *Abusing Science*, 52.
[108] Moore, *Realism and Christian Faith*, 50–2.
[109] Baehr, *The Inquiring Mind*.
[110] See Roberts and Wood, *Intellectual Virtues*, 3–31.

An example of such a vice is epistemic arrogance, which manifests itself in a disposition to draw illicit inferences regarding entitlements and exemptions whose consequence is the violation and erosion of the epistemic norms that regulate collective enquiry.[111] However, there is no suggestion that observing such epistemic virtues leads to explanatory success, in that such success (or failure) is contingent upon a complex range of factors, many of which lie beyond the knowledge or control of the epistemic agents involved. Instead, a more modest claim is made: namely, that the observance of such epistemic virtues will *typically* be conducive to enquiry.

The discussion in this chapter has raised two important questions, neither of which has yet been engaged: first, the question of what it means to 'explain' anything, whether at a scientific or theological level; and second, the question of the rational processes that are used in developing such explanations. In the next chapter, we shall give extended thought to the notion of rational explanation in science and religion.

---

[111] Roberts and Wood, *Intellectual Virtues*, 243–50.

5

Rational Explanation in Science and Religion

Rationality is often framed in terms of being able to explain our world—to be able to offer an account, however partial, of the interconnections of events and forces in our world that allow us to understand why certain things happen, or why they happen in a certain way.[1] Knowing *that* something has happened, or *that* something exists, is not the same as understanding *why* it happened, or *why* it exists. There is a significant gap between knowing *that* and knowing *why*. The belief that there is some reasonable explanation for what we observe within our world and experience within ourselves seems to be a universal human intuition; the task of finding such explanations—to make sense of things—is an integral aspect of the human engagement with reality, which so often focusses on 'why-questions'.[2] Yet the acceptance of this belief neither answers the question of what constitutes an 'explanation' nor reveals how the process of explanation actually works. It also remains unclear how the presumably narrower concept of 'scientific explanation' relates to explanation in general.[3]

Many theologians and philosophers of religion have also insisted that there is an explanatory function or dimension to religious belief. Basil Mitchell, for example, remarks that in its intellectual aspects,

---

[1] For some helpful discussions of this general theme, see Moore, 'Varieties of Sense-Making'; Achinstein, *The Nature of Explanation*; De Regt and Dieks, 'A Contextual Account of Scientific Understanding'; Friedman, 'Explanation and Scientific Understanding'; Newton-Smith, 'Explanation'.

[2] See Bromberger, 'Why-Questions'.

[3] Woody, 'Re-Orienting Discussions of Scientific Explanation'.

'traditional Christian theism may be regarded as a worldview or metaphysical system which is in competition with other such systems and must be judged by its capacity to make sense of all the available evidence'.[4] Others would be more cautious, not least in regard to what form of sense-making Christian theology is able to offer. It is one thing to confer meaning; it is quite another to offer an explanation of what happens in the world at large.

WHAT DOES IT MEAN TO 'EXPLAIN'?

For many, serious philosophical reflection on the notion of explanation began in 1948, with the publication of Carl G. Hempel and Paul Oppenheim's 'Studies in the Logic of Explanation'.[5] This work was intellectually generative, precipitating intense discussion of what it meant to 'explain' observations. By the beginning of the 1960s, there was growing interest in exploring the 'three basic components of a world picture' in order to achieve a maximum of 'explanatory coherence': observed objects and events, unobserved objects and events, and 'nomological connections' between them.[6] Although Hempel's 'deductive-nomological' model of scientific explanation was quickly shown to be incapable of encompassing the complexity of the natural sciences, it nevertheless stimulated intense discussion of alternative or supplementary accounts of explanation. The notion that explanation or understanding is about 'fitting phenomena into a comprehensive scientific world-picture' remains both attractive and defensible, even if there is more that needs to be said.[7]

---

[4] Mitchell, *The Justification of Religious Belief*, 99. Other Christian writers of the 1970s adopted similar positions: see, for example, Richmond, *Theology and Metaphysics*.

[5] Hempel and Oppenheim, 'Studies in the Logic of Explanation'. For reflections on its impact and subsequent critical discussion, see Salmon, *Four Decades of Scientific Explanation*, 11–50. The recent discovery of the text of Paul Feyerabend's 1948 essay 'Der Begriff der Verständlichkeit in der modernen Physik' offers some important illuminations of the development of this notion: see Feyerabend, 'Der Begriff der Verständlichkeit in der modernen Physik (1948)'.

[6] Sellars, *Science, Perception and Reality*, 356.

[7] For this phrase, see Salmon, *Causality and Explanation*, 77.

One of the happier results of this expanding interest in the concept of scientific explanation was the reappropriation and development of older approaches, most notably those of the great Victorian philosopher of science William Whewell (1794–1866). Whewell held that all observation involves what he terms 'unconscious inference', in that what is observed is actually unconsciously or automatically interpreted in terms of a set of ideas. Whewell rejected the somewhat deficient notion of induction as the mere enumeration of observations. Instead, Whewell developed the richer idea that induction was a process of reflection that added something essential to this process of enumeration—namely, some kind of organizing principle. In the process of induction, he suggested, 'there is a New Element added to the combination [of instances] by the very act of thought by which they were combined'.[8] Whewell held that this 'act of thought' was to be understood as a process of 'colligation'—the mental operation of bringing together a number of empirical facts by 'superinducing' upon them a way of thinking which unites the facts. For Whewell, this renders them capable of being expressed by a general law, which both identifies and illuminates the 'true bond of Unity by which the phenomena are held together'.[9]

Whewell's analysis leads to the conclusion that a good theory is able to 'colligate' observations that might hitherto have been regarded as disconnected, like a string holding together a group of pearls in a necklace. We might think, for example, of Newton's theory of gravity as 'colligating' observations that had up to that point been seen as unconnected—such as the falling of an apple to the ground, and the orbiting of planets around the sun. This idea of explanation as colligation of what might otherwise be seen as unrelated and disparate events underlies the notion of unitative explanation, which we shall consider presently (132–5).[10]

There remains a significant divergence between the natural sciences and social sciences on how 'explanation' is to be understood. This is often framed in terms of the distinction between *Verstehen*

---

[8] Whewell, *Philosophy of the Inductive Sciences*, vol. 2, 48. For Whewell's approach and its wider impact, see Snyder, *Reforming Philosophy*, 33–94.
[9] Whewell, *Philosophy of the Inductive Sciences*, vol. 2, 46.
[10] e.g. Morrison, *Unifying Scientific Theories*.

(an interpretative understanding) and *Erklären* (a law-governed explanation).[11] *Erklären*—the dominant approach to explanation in the natural sciences—aims to make explanatory sense of an observed phenomenon by identifying the laws that govern it, whereas *Verstehen* represents an attempt to make empathetic sense of the phenomenon by looking for the perspective in which the phenomenon appears to be meaningful and appropriate. The social sciences tend to the view that aiming for a merely explanatory sense of social phenomena makes it impossible to allow comprehensive knowledge of these phenomena, in that for such knowledge the observer has to come to terms with the perceived significance of these events within a social context. An investigator is thus studying not passive objects, but active interpretative agents—how people understand their world, and how that understanding shapes their practice.

Yet the intense discussion of what it means to 'explain' something in the natural sciences has not led to the clarification that many had hoped for. A recent review of debates about the nature of scientific explanation suggests that fifty years of intense discussion have led not to the emergence of a consensus, but, on the contrary, to an increasingly diverse account of the nature of explanation. It is as if a 'metatheory' is required—'some deeper theory that explained what it was about each of these apparently diverse forms of explanation that makes them explanatory'.[12] The present situation, characterized by the absence of any such theory, is seen by some influential writers as 'an embarrassment for the philosophy of science'. The question of what it means to explain something still remains to be resolved.

Important questions remain about the limits of explanation. In an important recent discussion of the explanatory capacities of the natural sciences, Steven Weinberg remarks that it seems clear that 'we will never be able to explain our most fundamental scientific principles'.[13] We can certainly hope to develop a 'set of simple universal laws of nature'; these, however, cannot be explained. They exist; they have explanatory capacity, on the basis of accepted models

---

[11] Apel, 'The *Erklären-Verstehen* Controversy in the Philosophy of the Natural and Human Sciences'.
[12] Newton-Smith, 'Explanation', 130–2.
[13] Weinberg, 'Can Science Explain Everything?', 36.

of scientific explanation; yet they themselves cannot be explained. An explanatory regress thus terminates at a certain point.

Cognitive science casts some helpful light on the processes by which the human brain makes sense of our world.[14] The human mind has evolved many cognitive tools for this purpose including abstraction, counterfactual thought, deduction, and induction. One of the most important of these processes is causal thinking. An individual's ability to determine if a precipitating event was the cause of any given outcome is essential for making sense of the complex world in which we live. At present, there are four significant psychological models of causal inference, none of which appears adequate to deal with all aspects of the phenomenon. Yet it remains unclear what it means to 'explain' something—a difficulty which is compounded by the multiple meanings of the term 'explain' in the first place.[15]

These deep-seated and seemingly intractable concerns help to frame our discussion of the place of explanation in science and in Christian theology. There are some—such as Christopher Hitchens, one of the leading representatives of the 'New Atheism'—who hold that religion is incapable of explaining anything, and see this as an epistemic vice. However, the rhetorical force of this bold assertion is significantly reduced by Hitchens's striking failure to clarify what he means by 'explain', which, of course, echoes a wider problem within philosophy in general, and the philosophy of science in particular.

Others similarly argue that religion is unable to explain anything—but see this distinguishing feature as neutral, if not even a virtue. Terry Eagleton, for example, is severely critical of those who treat religion as a fundamentally explanatory phenomenon. 'Christianity was never meant to be an explanation of anything in the first place', he argued. 'It's rather like saying that thanks to the electric toaster we can forget about Chekhov.'[16] Eagleton suggests that believing that religion is a 'botched attempt to explain the world' is about as helpful as 'seeing ballet as a botched attempt to run for a bus'. There are

---

[14] See Fugelsang and Dunbar, 'A Cognitive Neuroscience Framework for Understanding Causal Reasoning and the Law'; Operskalski and Barbey, 'Cognitive Neuroscience of Causal Reasoning'.
[15] Șerban, 'What Can Polysemy Tell Us About Theories of Explanation?'
[16] Eagleton, *Reason, Faith, and Revolution*, 7.

parallels here with the account of science offered by the philosopher Bas van Fraassen, who rejects as a 'false ideal' any notion that 'explanation is the *summum bonum* and exact aim of science',[17] or those who hold that human 'understanding' is a purely subjective phenomenon that should not be allowed to play a role in the epistemic evaluation of scientific theories and explanations.[18] Yet this pragmatic approach to scientific understanding is open to criticism, not least because understanding is to be seen as an essential element of the epistemic aims of science.

While Eagleton is surely right to argue that there is more to Christianity than an attempt to make sense of things, others would argue that some kind of explanatory capacity is an integral—though not necessarily a fundamental or central—theme of the Christian faith.[19] Some critics of religion follow Frazer's flawed account of religion as a kind of primitive cosmology, which competes (unsuccessfully) with the natural sciences as an explanation of reality.[20] Yet as I shall indicate in this chapter, the current understanding of the notion of 'explanation' in the natural sciences is actually such that religion can be said to be capable of 'explaining' our world in its own distinct terms.

To develop the way in which the notion of 'making sense' or 'explaining' is currently understood, we shall focus on the two most widely discussed approaches to scientific explanation, which focus on the ideas of causation and unification respectively.

CAUSALITY AS EXPLANATION

If *A* causes *B*, then *A* explains *B*. This line of thought has played a significant role in reflection on the nature of explanation, and is seen

---

[17] van Fraassen, 'The Pragmatics of Explanation', 144.
[18] Trout, 'Scientific Explanation and the Sense of Understanding'.
[19] See the nuanced position developed in Plantinga, *Knowledge and Christian Belief*.
[20] Dawkins, *The God Delusion*, 188. For criticisms of this, see Crane, *The Meaning of Belief*, ix–xiii. Crane presents religion almost entirely as an intellectual-rational construct, which seriously limits his appreciation of its imaginative and emotional dimensions. For a much better account, see Smith, *Religion*, 20–76.

by many as the most intuitive approach. According to Wesley Salmon, a causal theory of explanation is appropriate because 'underlying causal mechanisms hold the key to our understanding of the world'.[21] Amplifying this point, Salmon argues that 'causal processes, causal interactions, and causal laws provide the mechanisms by which the world works; to understand why certain things happen, we need to see how they are produced by these mechanisms'.[22] This type of approach is open to development in a number of ways, particularly through the use of Bayesian theory and related paradigms. Causal patterns may be distinguished from mere correlation by construction of 'causal Bayes nets', which identify patterns of nested contingencies in a manner that allows non-causal relations to be filtered out.[23] Furthermore, we can assert that A causes B without needing to know what caused A.[24]

Neither Salmon nor any recent commentator on causal explanations notes what may well be one of the most important points in its favour—that causal reasoning can be argued to confer evolutionary advantages. In their evolutionary history, humans have had to deal with the threat of predators. Learning to detect agency behind natural phenomena and events around us could thus confer survival potential. This has led some to suppose (though this belief remains unevidenced) that humans have evolved a hypersensitivity in detecting intentional agents at a perceptual level, based on the notion that failing to detect these causal agents may potentially be more harmful than incorrectly assuming that agents are absent.[25] It remains far from clear, however, whether this inference is justified.[26]

Yet problems remain with causal explanations, especially within the field of quantum theory. Among the most often cited are the

---

[21] Salmon, *Scientific Explanation and the Causal Structure of the World*, 260. See further the detailed discussion and original proposals in Woodward, *Making Things Happen*.
[22] Salmon, *Scientific Explanation and the Causal Structure of the World*, 132.
[23] Gebharter, 'Causal Exclusion and Causal Bayes Nets'.
[24] Lipton, 'What Good Is an Explanation?', 8.
[25] Boyer, *Religion Explained*; McCauley, *Why Religion Is Natural and Science Is Not*.
[26] See Maij, van Schie, and van Elk. 'The Boundary Conditions of the Hypersensitive Agency Detection Device'.

Einstein-Podolsky-Rosen (EPR) correlations, which cannot be explained by means of direct causal influence or by referring to a common cause. Salmon was quite clear that his causal-mechanical model of explanation could not cope with this specific case.[27] Some would argue that this difficulty is only to be expected, in that 'quantum physics is only indirectly a science of reality but more immediately a science of knowledge'.[28] The significance of this problem might, of course, be deflected by pointing out that any version of quantum theory is strongly interpretative; or by appealing to forms of quantum mechanics (such as those developed by Louis de Broglie and David Bohm) which are framed in causative terms. Yet although there is a clear diversity of preference within the field of quantum mechanics, these two approaches are regarded by most as inferior options.[29] Underlying this is perhaps something more fundamental: the absence of a generalized account of causation in the first place. Hume's concerns about the notion have never been properly resolved; although many solutions have been proposed, none has secured widespread acceptance.[30]

A further issue that needs to be noted here is the interconnection between physical causation and explanation. Which is prior? Philip Kitcher argued that 'the "because" of causation is always derivative from the "because" of explanation'.[31] In other words, our causal judgements merely echo the explanatory relationships that result from our attempts to construct unified theories of nature. There is thus no independent causal order which our explanations must capture or represent.

A final concern that needs to be raised relates to the notion of mathematical explanation.[32] The 'unreasonable effectiveness' of

---

[27] Unfortunately, Woodward does not engage with quantum theory in general, or this specific topic, describing it as a 'recherché case': Woodward, *Making Things Happen*, 92.

[28] Brukner and Zeilinger, 'Quantum Physics as a Science of Information', 47–8.

[29] Brown and Wallace, 'Solving the Measurement Problem'. For the diversity of approaches, see Schlosshauer, Kofler, and Zeilinger, 'A Snapshot of Foundational Attitudes toward Quantum Mechanics'.

[30] One of the most interesting is Dowe, *Physical Causation*. For others, see Sosa and Tooley, *Causation*.

[31] Kitcher, 'Explanatory Unification and the Causal Structure of the World', 477.

[32] On which see Barrow, 'Mathematical Explanation'.

mathematics in being able to represent some fundamental features of the universe is well known.³³ When scientists try to make sense of the complexities of our world, they use 'mathematics as their torch'. It is far from clear why this is the case. 'The miracle of the appropriateness of the language of mathematics to the formulation of the laws of physics is a wonderful gift which we neither understand nor deserve'.³⁴ Yet mathematics is able to achieve this explanatory function without being causative in any meaningful sense of the word. It seems that causation is best seen as one explanatory strategy among others; not all causes are explanatory, nor are all explanations causal in character.

UNIFICATION AS EXPLANATION

The most influential alternative to the causal-mechanical approach is at present the unificationist conception of explanatory understanding, which holds that science achieves explanation by uncovering a unified picture of the world. A scientific explanation can thus be understood as providing a unified account of a range of different phenomena. To understand any given phenomenon is to see how it fits together with other phenomena within a unified whole, discerning the fundamental unity that underlies the apparent diversity of the phenomena themselves.³⁵

Historically, it is easy to show that theory unification has clearly played an important role in the development of the natural sciences. Paradigmatic examples include René Descartes's unification of algebra and geometry, Isaac Newton's unification of terrestrial and celestial theories of motion, and James Clerk Maxwell's unification of electricity and magnetism.³⁶ Yet this is simply one form of unification, in

[33] Wigner, 'The Unreasonable Effectiveness of Mathematics'.
[34] Wigner, 'The Unreasonable Effectiveness of Mathematics', 9.
[35] See Bartelborth, 'Explanatory Unification'; Plutynski, 'Explanatory Unification and the Early Synthesis'; Schweder, 'A Defense of a Unificationist Theory of Explanation'.
[36] On which see Grosholz, 'Descartes' Unification of Algebra and Geometry'; Morrison, 'A Study in Theory Unification'; eadem, *Unifying Scientific Theories*, 4.

which phenomena that were previously regarded as having quite different causes or explanations were realized to be the outcome of a common set of mechanisms or causal relationships—as in Newton's demonstration that the orbits of the planets and the motion of terrestrial objects falling freely close to the surface of the earth are actually due to the same force of gravity and obey the same laws of motion.

To understand the point at issue, we may reflect on the transition from classical Newtonian mechanics to relativistic quantum mechanics, which took place during the first few decades of the twentieth century.[37] In the eighteenth and nineteenth centuries, classical mechanics was seen as a self-sufficient and intellectually autonomous area of theory, capable of accounting for what could be observed in nature. Based on the extensive earlier observational and analytical work of individuals such as Nicolas Copernicus, Johann Kepler, Galileo Galilei, and Isaac Newton, classical mechanics was widely regarded as a fundamental theory, capable of mathematical formalization.[38] It did not need to be positioned within a richer intellectual framework to be understood, but was an autonomous and essentially complete theory.

Yet following the work of Max Planck, Albert Einstein, and Niels Bohr in the early twentieth century, it was realized that classical mechanics was a special, limiting case of a more complete theory. As the theory of quantum mechanics developed in response to a growing body of evidence which older theoretical models simply could not accommodate, it became clear that relativistic quantum mechanics was the more fundamental theory, capable of far greater explanatory capacity. And, perhaps most importantly of all for our purposes, this more fundamental theory was able to account for both the successes and the failures of classical mechanics, by identifying its limited sphere of validity. The 'correspondence principle', first identified by Niels Bohr in 1923, sets out, clearly and elegantly, how quantum mechanics reduces to classical mechanics under certain limits.[39]

---

[37] This topic is covered in standard textbooks, such as Tang, *Fundamentals of Quantum Mechanics*.
[38] Illiffe, 'Newton, God, and the Mathematics of the Two Books'.
[39] For comment, see Pais, *Niels Bohr's Times, in Physics, Philosophy and Polity*, 192–6.

Neither relativistic quantum mechanics nor quantum field theory invalidated classical mechanics; they showed that it was rather a special case of a more comprehensive and complex theory.[40] The classical model was thus accounted for on the basis of the greater explanatory capacity of the relativistic model, and the limits of their correspondence established. And, perhaps more importantly, the relativistic approach explained why the classical theory worked in certain situations, and not in others. Its validity was affirmed within specific limits.

We see here an important and well-understood insight from the world of scientific theory development: that a better theory is able to accommodate all the valid insights of an earlier theory, while at the same time expanding its horizons and identifying the basis of its plausibility.[41] A theory with considerable explanatory capacity is able to create conceptual space, valid under certain limiting yet significant conditions, for a theory which might, at first sight, appear to be quite independent, yet, on closer examination, turns out to be a special case of the higher-order theory, which provides a unifying framework which colligates and correlates theories that might have been seen to be inconsistent or incoherent.

Yet there are other forms of unification, such as the creation of a common classificatory scheme or descriptive vocabulary where no such scheme previously existed (such as Linnaeus' comprehensive system of biological classification), or the creation of a common mathematical framework or formalism which can be applied to many different classes of phenomena. In 1989, Philip Kitcher presented an early formulation of the unificationist approach, which emphasized the importance of discerning common patterns within nature as the basis of explanation:[42]

Understanding the phenomena is not simply a matter of reducing the 'fundamental incomprehensibilities' but of seeing connections, common

---

[40] See, for example, Landsman, *Mathematical Topics between Classical and Quantum Mechanics*, 7–10.

[41] For 'progress as incorporation' in the development of scientific theories, see John Losee, *Theories of Scientific Progress: An Introduction*. New York: Routledge, 2004, 5–61.

[42] Kitcher, 'Explanatory Unification and the Causal Structure of the World', 432.

patterns, in what initially appeared to be different situations.... Science advances our understanding of nature by showing us how to derive descriptions of many phenomena, using the same patterns of derivation again and again, and, in demonstrating this, it teaches us how to reduce the number of types of facts we have to accept as ultimate (or brute).

Unificationist approaches to scientific explanation hold that successful explanations unify our knowledge of the world, in that they allow what might otherwise be seen as disparate and disconnected observations to be seen as aspects of a greater overall explanatory pattern. The unifying power of a scientific theory could thus be argued to increase in proportion to its generality, its simplicity, and its cohesion. Explanations contribute to our understanding of the world by showing how phenomena can be embedded within general nomic patterns that we recognize in the world. Explanation is thus about the articulation of a theory's capacity to systematize or render a set of observations coherent, by allowing them to be seen within a specific informing context.

TWO APPROACHES TO EXPLANATION: ONTIC AND EPISTEMIC

Some philosophers have argued that these two accounts of explanation illustrate two distinct—yet potentially complementary—approaches to our understanding of how the natural sciences work: 'epistemic' and 'ontic'. This distinction is due to Wesley Salmon, who argued that ontic accounts of explanation hold that explanations involve the identification of those ontic structures in the world which are responsible for the production of the phenomena that are to be explained, whereas epistemic accounts hold that explanation is concerned with making phenomena understandable, predictable, or intelligible by setting them in an informing context.[43] Although the manner in which the categories of 'ontic' and 'epistemic' explanations

---

[43] Salmon, *Causality and Explanation*; idem, *Four Decades of Scientific Explanation*. For a good summary and evaluation of Salmon's ontic approach, see Wright, 'The Ontic Conception of Scientific Explanation'.

are framed has since shifted slightly,[44] this dichotomy has proved helpful in exploring different approaches to explanation, and how they might relate to one another.

Epistemic accounts assign an essential role to the conceptual structures by which an explanation is conveyed, and tend to understand explanation as a cognitive achievement. Hempel's older Deductive-Nomological approach falls into this category, as does the unificationist approach. Here, an explanation is fundamentally seen as demonstrating how certain observations or existing theories can be subsumed under, or incorporated within, a wider generalization.

Causal accounts of explanation, however, are ontic, holding that an explanation identifies the causal process or mechanism that gives rise to what is observed.[45] Explanation is thus about the exploration of the metaphysics and epistemology of mechanisms.[46] Given the respective merits of both approaches, it should be no cause for surprise that some philosophers have tried to develop an approach to explanation that merges aspects of unificationist and causal models.[47] Such accounts of explanation attempt to reconcile the ontic and epistemic approaches to explanation, arguing that the best explanations fulfil epistemic and ontic norms and roles simultaneously. It could, for example, be argued that a unitative explanation of a set of observations cannot be fully successful unless it is fundamentally constrained to be grounded in a true or accurate view of the world, and getting things right in this way is not itself fundamentally constrained or determined by any epistemic norms.[48] This view thus asserts the normative priority of ontic norms, but not their exclusivity. Yet there is a growing awareness that no single account of explanation is, in the first place, adequate, and in the second, superior. We may have to come to terms with such approaches simply being different,[49] and trying to correlate them as best we can.

---

[44] Illari, 'Mechanistic Explanation: Integrating the Ontic and Epistemic'.
[45] Craver, 'The Ontic Account of Scientific Explanation'.
[46] See especially Machamer, 'Activities and Causation'.
[47] Such as Strevens, 'The Causal and Unification Approaches to Explanation Unified'; Bangu, 'Scientific Explanation and Understanding: Unificationism Reconsidered'.
[48] Craver and Kaiser, 'Mechanisms and Laws: Clarifying the Debate'.
[49] Craver, 'The Ontic Account of Scientific Explanation', 35.

Yet any distinction between epistemic and ontic accounts of causality is fuzzy, partly because of the entanglement of theory and observation. Many would suggest that an epistemic framework is unconsciously or covertly deployed in making ontic judgements about causality. Inferring any 'causal structures' is thus a matter of faith, resting on certain judgements and interpretations.[50] It is well known that such a reasoning process can go awry, so that causality may be 'observed' or inferred when there is no causal process in the first place. For example, some argue that belief in gods arises from the perception of agency or causality, so that natural events are interpreted as caused by supernatural agents.[51]

RELIGIOUS EXPLANATION: SOME GENERAL REFLECTIONS

We have already noted that some theologians and philosophers of religion take an essentially Wittgensteinian approach to its explanatory function, seeing religious statements as meaningless (*Bedeutungs los*) and nonsense (*Unsinn*).[52] Wittgenstein's position is actually more nuanced than this influential interpretation suggests. To begin with, Wittgenstein had concerns about the 'meaningfulness' of *any* attempt to represent the world philosophically.[53] Furthermore, his provocative suggestion that philosophical problems might be seen 'from a religious point of view' did not mean that he conceived them as *religious* problems, but rather that he discerned 'a similarity, or

---

[50] Sloman, *Causal Models*, 46–9; 101–14. See also Shadish, Cook, and Campbell, *Experimental and Quasi-Experimental Designs for Generalized Causal Inference*, 1–32.
[51] Barrett, *Why Would Anyone Believe in God?*; Maij, van Schie, and van Elk, 'The Boundary Conditions of the Hypersensitive Agency Detection Device'.
[52] For a particularly influential statement of this position, see Phillips, *Religion without Explanation*. For criticism, see van Holten, 'Does Religion Explain Anything?'
[53] Note especially Wittgenstein, *Tractatus Logico-Philosophicus*, §4.003: 'Die meisten Sätze und Fragen, welche über philosophische Dingen geschrieben worden sind, sind nicht falsch, sondern unsinnig.'

similarities, between his conception of philosophy and something that is characteristic of religious thinking'.[54]

Most theologians, however, would argue that Christianity does offer an explanatory framework,[55] even if this is not its primary focus, often using visual imagery to develop this theme. The French philosopher Simone Weil (1909-43) is a good recent example of this approach:[56]

> If I light an electric torch at night out of doors, I don't judge its power by looking at the bulb, but by seeing how many objects it lights up. The brightness of a source of light is appreciated by the illumination it projects upon non-luminous objects. The value of a religious or, more generally, a spiritual way of life is appreciated by the amount of illumination thrown upon the things of this world.

The ability to illuminate reality is an important measure of the reliability of a theory, and an indicator of its truth. This perception underlies Karl Popper's 'searchlight' approach to scientific theory generation, in which theories or hypotheses are developed to ascertain how well they illuminate the world of observation.[57]

The philosopher Keith Yandell offers a representative account of this explanatory or sense-making aspect of religion:[58]

> A religion is a conceptual system that provides an interpretation of the world and the place of human beings in it, bases an account of how life should be lived given that interpretation, and expresses this interpretation and lifestyle in a set of rituals, institutions and practices.

Now it could easily be objected that such explanatory frameworks are found beyond the category of religion—for example, in Marxism, or the 'universal Darwinism' of Richard Dawkins.[59] Nevertheless, while

---

[54] Malcolm, *Wittgenstein: A Religious Point of View?*, 24.

[55] For a good survey of approaches, see Dawes, *Theism and Explanation*.

[56] Simone Weil, *First and Last Notebooks*. London: Oxford University Press, 1970, 147.

[57] Popper first seems to use this approach in his unpublished PhD thesis *Zur Methodenfrage der Denkpsychologie* (1928). See ter Hark, 'Searching for the Searchlight Theory', 466. For a more popular application of this 'searchlight' approach, see Pennock, *Tower of Babel*, 53-4.

[58] Yandell, *Philosophy of Religion*, 16.

[59] On this 'universal Darwinism', see Dawkins, *A Devil's Chaplain*, 78-90.

this explanatory capacity may not be a *distinguishing* feature of religion, setting it apart from everything else, it can certainly be argued to be *characteristic* of it.

It is the Christian vision of reality *as a whole*—rather than any of its individual components—that proves imaginatively and rationally compelling. Individual observations of nature do not 'prove' Christianity to be true; rather, Christianity validates itself by its ability to make sense of those observations. As W. V. O. Quine argues in his 'Two Dogmas of Empiricism', what really matters is the ability of a theory *as a whole* to make sense of the world. Our beliefs are linked in an interconnected web that relates to sensory experience at its boundaries, not at its core.[60] The only valid test of a belief, Quine argued, is thus whether our experience fits into an overall interconnected web of beliefs.[61]

This is an important element of C. S. Lewis's theistic account of religious explanation, which is often summed up in his signature statement from an Oxford lecture of 1945: 'I believe in Christianity as I believe that the Sun has risen, not only because I see it, but because by it I see everything else.'[62] For Lewis, belief in God is both evidenced and evidencing. There are hints here of the self-evidencing explanations noted by Carl Hempel, in which what is explained constitutes an important element of our reasons for believing that the explanation itself is correct.[63] This may be set alongside Wittgenstein's observation that one and the same proposition or idea may at one point be treated as something that is *to be tested*, and at another as a *rule of testing*.[64]

Peter Lipton provides an example of this circularity drawn from modern cosmology, noting that the velocity of recession of a galaxy is held to explain the red shift of its characteristic spectrum, even if the observation of that shift is itself an essential part of the scientific

---

[60] Quine's argument is weakened at this point by his ahistorical and purely logical approach to the topic, which contrasts sharply with that of Pierre Duhem: see Pietsch, 'Defending Underdetermination or Why the Historical Perspective Makes a Difference'.

[61] Quine, 'Two Dogmas of Empiricism'.

[62] Lewis, *Essay Collection*, 21.

[63] Hempel, *Aspects of Scientific Explanation*, 370–4.

[64] Wittgenstein, *On Certainty*, 98.

evidence that the galaxy is indeed receding at that the specified velocity.[65] Self-evidencing explanations thus exhibit a kind of circularity, which can be set out as follows: A explains B while B justifies A. There is nothing invalid or improper about this form of explanatory argument, which is widely encountered in scientific explanation. Good explanations may be self-evidencing, even if this appears to involve at least some degree of self-referential circularity. For Lewis, Christian theology is to be seen as intellectually capacious, capable of accommodating the complexities of observation and experience, and is to be judged partly by that capacity.

A more theologically sophisticated account of this approach was developed by the philosopher of religion Ian T. Ramsey in 1964. For Ramsey, the best way of evaluating the correspondence between theology and experience or observation was by trying to assess the degree of 'empirical fit' between the theory and the empirical world.[66]

> The theological model works more like the fitting of a boot or a shoe than like the "yes" or "no" of a roll call. In other words, we have a particular doctrine which, like a preferred and selected shoe, starts by appearing to meet our empirical needs. But on closer fitting to the phenomena the shoe may pinch... The test of a shoe is measured by its ability to match a wide range of phenomena, by its overall success in meeting a variety of needs. Here is what I might call the method of empirical fit which is displayed by theological theorizing.

RELIGIOUS EXPLANATION: ONTIC AND EPISTEMIC

William Lycan, in reflecting on human judgement and justification, insisted that 'all justified reasoning is fundamentally explanatory reasoning that aims at maximizing the "explanatory coherence" of one's total belief system'.[67] The question of how a belief system is derived is seen as secondary; the critical question is its capacity to make sense of things. Beliefs are to be judged 'through the roles they

---

[65] Lipton, *Inference to the Best Explanation*, 24.
[66] Ramsey, *Models and Mystery*, 17. See further Tilley, 'Ian Ramsey and Empirical Fit'.
[67] Lycan, *Judgement and Justification*, 128.

play in a maximally coherent explanatory system and not because of anything in particular to do with the mechanisms that produced them'.[68] This statement needs to be set alongside a wider discussion about a 'logic of discovery' and a 'logic of justification', noted earlier (159–64).

Religious explanation can be set within this informing context, while still bearing a clear family resemblance to other modes of knowledge generation. Ernst Sosa offered an account of justification of beliefs which, while having widespread applicability, is of particular importance for theological rationality: 'A belief is justified only if it fits coherently within the epistemic perspective of the believer.'[69] For Sosa, the justification of any of our beliefs *always* takes place within an epistemic perspective.

Christianity gives rise to such epistemic perspectives—yet with nested or embedded ontic elements. The theological frameworks set out by Augustine of Hippo, Athanasius of Alexandria, Blaise Pascal, and C. S. Lewis all—though in their different ways—set out a core belief that a rational God created a coherent and rational (*logikos*) universe, whose structures reflect the character of its creator, and are capable of being grasped by the human mind, and their significance appreciated and represented, if only dimly and partially. This framework of belief incorporates both ontic and epistemic elements. In the first place, it affirms that the being of the universe ultimately derives from the being of God; in the second, it affirms that humanity, in bearing the image of God, has a created capacity to engage, interpret, and understand the universe.[70] More specifically, it affirms that the Christian 'belief system' is capable of accommodating our observations and experiences within an intelligible and coherent greater whole.[71]

This general theme is predominant in the thought of C. S. Lewis, and is best seen in his 'argument from desire,' as set out in *Mere*

---

[68] Lycan, *Judgement and Justification*, 172–3.
[69] Sosa, *Knowledge in Perspective*, 145.
[70] See here especially Peters, *The Logic of the Heart*. For the relation of the ontic and noetic dimensions of Barth's doctrine of creation, see Gabriel, *Barth's Doctrine of Creation*, 15–16.
[71] McGrath, *Re-Imagining Nature*, 41–68.

Christianity.[72] As noted earlier (105), Lewis's *explanandum* lies in the domain of human experience: many long for something of ultimate significance, only to find their hopes dashed and frustrated when they attain it. 'There was something we grasped at, in that first moment of longing, which just fades away in the reality.'[73] So what does this experience mean? What interpretative framework explains this experience, and offers an indication of its origins and goal?

Lewis's argument is often set out in the following way:

1. Every natural or innate desire in us corresponds to a real object that can satisfy the desire.
2. There exists a desire within us that nothing created or finite can satisfy.
3. There thus exists something beyond the realms of the created and finite which can satisfy this desire.

A closer examination of Lewis's argument, however, suggests that it is actually framed using what Lewis himself termed 'supposals'. *Suppose* there is a God who created us to relate to him, as Christianity affirms to be the case.[74] This provides an intellectual framework which explains why only God is able to fulfil the deepest longings of humanity, so that we will never be satisfied by anything that is created or finite. Does not this framework fit in well with what we actually experience of reality? And is not this resonance of supposal and observation indicative of the truth of the supposal in the first place? This is the style of thinking, bearing a strong resemblance to what Peirce styles 'abduction', which leads to his conclusion that 'if I find in myself a desire which no experience in this world can satisfy, the most probable explanation is that I was made for another world'.[75] The structure of Lewis's argument can thus be set out as follows.

---

[72] McGrath, 'Arrows of Joy: Lewis's Argument from Desire'.
[73] Lewis, *Mere Christianity*, 135.
[74] For this idea of a 'supposal', see Lewis's letters to Mrs Hook, 29 December 1958; *The Collected Letters of C. S. Lewis*, edited by Walter Hooper, 3 vols. San Francisco: HarperOne, 2004–6, vol. 3, 1004–5; and to Sophia Storr, 24 December 1959; *Letters*, vol. 3, 1113–14. 'Supposing there was a world like Narnia, and supposing, like ours, it needed redemption, let us imagine what sort of Incarnation and Redemption Christ would have there' (1113).
[75] Lewis, *Mere Christianity*, 136–7.

1. We experience a 'desire which no experience in this world can satisfy'.
2. Suppose that the Christian way of thinking is right, and that this desire is interpreted within its framework; these resonances or harmonies with experience would then be expected.
3. Therefore there are additional grounds for holding that the belief system of Christianity is true.

Lewis's approach thus melds both ontic ('this is the way things are') and epistemic ('this enables us to understand things') elements, holding them both to be true, while leaving open the question of which is to be given priority or privilege. In practice, Lewis's trajectory of thought is from the ontic to the epistemic: in other words, the structure of the world, when seen from a Christian perspective, leads to a certain way of thinking about the world—or, to remain more faithful to Lewis's own way of speaking, a certain way of *seeing* reality. Lewis thus adopts what comes close to a unificationist approach to explanation, while nevertheless grounding this in an ontic account of reality.

Other theologians would speak more explicitly about a distinct 'Trinitarian logic' of the Christian faith,[76] distinguishing this from theistic or deistic alternatives, and seeing this as having potential explanatory capacity exceeding those offered by philosophers of religion, who tend to avoid discussion of Trinitarian conceptions of God, or reflection on their potential to illuminate the world.[77]

THEOLOGY, ONTOLOGY, AND EXPLANATION

Earlier, we noted epistemic approaches to explanation which seeks to offer a unified account of scientific theories about the world. So do

---

[76] See, for example, Polkinghorne, 'Physics and Metaphysics in a Trinitarian Perspective'; Torrance, *God and Rationality*, 3–27; Lemeni, 'The Rationality of the World and Human Reason as Expressed in the Theology of Father Dumitru Stăniloae'.

[77] For examples, see Dawes, *Theism and Explanation*; O'Connor, *Theism and Ultimate Explanation*. O'Connor's carefully constructed case for identifying a 'necessary being' with a 'transcendent, personal creator' (*Theism and Ultimate Explanation*, 86–110, 130–42) establishes an affinity, but not an absolute identity, with a Christian understanding of God.

these successful unifications simply relate to formalizations of nature, or might they be seen as pointing to the ontological unity of nature, so that a 'successful phenomenological theory' can be seen as evidence for 'an ontological interpretation of theoretical parameters'?[78] This points to the fundamental source of explanatory power lying in *ontology*—an understanding of the way things are, of the fundamental order of things—and thus suggests that there is, as most scholars now seem to accept, an irreducible ontic element in the process of explanation. Pierre Duhem (1861-1916) thus argued that to explain something 'is to strip reality of the appearances covering it like veils, in order to see this naked reality face to face'.[79] It is by discovering the 'big picture' that its individual elements are able to be both known and understood; explanation is about the location of an event or observation within this deeper context.

Classical Christian theology develops such a vision of reality, which it believes to have both explanatory capacity and virtue. It is not a form of 'onto-theology', based on a priori first principles, but rather emerges from wrestling with the foundational events of the Christian faith. Nor is this explanatory capacity to be seen as the *primary* aspect of religious faith, even though the ability to make sense of the world and identify patterns of meaning is an important element of any religious system, and serves an important psychological function.[80] The early Pauline letters of the New Testament, for example, focus on the theme of salvation in Christ, which is affirmed as a universal possibility, extending beyond the people of Israel. Yet even within the New Testament, an emergent Trinitarianism can be discerned, both as a framework that arises from the narrative of Jesus Christ and as a framework which interprets this narrative.[81] Trinitarianism is a good example of the kind of self-evidencing explanations noted by Carl Hempel, in which the quality of what is explained constitutes a ground for believing that the explanation itself is correct (139–40).

---

[78] Morrison, *Unifying Scientific Theories*, 108.

[79] Duhem *La théorie physique*, 3-4. 'Expliquer, *explicare*, c'est dépouiller la réalité des apparences qui l'enveloppent comme des voiles, afin de voir cette réalité nue et face à face.'

[80] Wolf, *Meaning in Life*; Wong, *The Human Quest for Meaning*.

[81] For the development and contemporary interpretation of this doctrine, see Woźniak and Maspero, eds, *Rethinking Trinitarian Theology*.

Christians believe in the Trinity, partly because they believe it arises naturally from the narratives and emerging doctrinal statements of the New Testament, and partly because this framework provides a lens which interprets their experiences and observations. Such a framework, for example, can be argued to be presupposed in Julian of Norwich's attempt to make sense of her complex experiences, allowing her to make the transition from *what* she saw to *who* was showing it to her, and appreciating its implications.[82] Such a framework enables a response to the human experience of suffering which safeguards the transcendence of God, while at the same time allowing such suffering to be seen in a way that is both meaningful and generative.[83] In each of these two examples, the doctrine of the Trinity functions as a *theoria*, a way of beholding things, allowing us to see our world in a new way, and also to understand our experience in a new way.[84] It provides an informing perspective, an interpretative framework, which enables us to discern how events and experiences fit into broader patterns. It thus offers primarily, yet not exclusively, an epistemic model of explanation, by throwing a conceptual net over the complexities of experience, so that these may both be captured and colligated. To understand something is to locate it within a web of meaning.

A CASE STUDY: AQUINAS'S 'SECOND WAY'

This basic approach is found in Thomas Aquinas's 'Five Ways', often misleadingly described as 'proofs of God's existence'. Traditionally, many writers have sought to demonstrate the intrinsic rationality of arguments for the existence of God. Although some take these to prove God's existence, these lines of argument are really affirmations of the fundamental rationality of Christian belief.[85] Such arguments,

---

[82] Watson, 'The Trinitarian Hermeneutic in Julian of Norwich's *Revelation of Divine Love*', 67–72.
[83] Coman, 'Suffering in the Trinitarian Pattern of Redemption'.
[84] A theme developed at several points by the German theologian Eberhard Jüngel. See especially Jüngel, *Erfahrungen mit der Erfahrung*.
[85] Davis, *God, Reason and Theistic Proofs*, 189–90.

whether ontic or epistemic, are generally deployed to show that theists are rational in their belief in the existence of God, although the strategies developed range across a range of options, including: arguing that it is more rational to believe that God exists than it is to deny that God exists; demonstrating that it is more rational to believe that God exists than to be agnostic on the existence of God; or that it is as rational to believe in God as it is to believe in many of the things that atheist philosophers often believe in (such as the existence of 'other minds' or the objectivity of moral right and wrong). Aquinas's approach is of especial interest, on account of the explicit connection he proposes between the existence of God and the human capacity to make sense of our world.

For Aquinas, God can be considered to be an explanatory agent, whose existence and nature provide a retrospective explanation of various aspects of our experience of the world—such as the ordering of the world, or our sense of goodness or beauty. In the 1948 debate between Bertrand Russell and Frederick Copleston, Russell ruled out there being any explanation of the universe. 'The universe is just there, and that's all.'[86] Aquinas, however, takes the view that it is reasonable to seek an explanation of why the world exists, and why it has its distinct characteristics. The Universe is seen to require 'an explanation in terms of a relationship to something other than itself'—that is, God.[87] A similar point is made by Thomas Nagel, who argues that the existence of our universe requires a larger explanatory context than scientific laws, in that such a limited explanation 'would still have to refer to features of some larger reality that contained or gave rise to it'.[88]

Aquinas's approach to explanation is generally causal, perhaps even mechanical. Consider the structure of the second of the 'Five Ways,' the *ratio causae efficientis*, which can be set out in four stages.[89]

1. We observe an order of efficient causes in the things we see around us.
2. Yet we do not observe, and cannot expect to observe, anything that is the efficient cause of itself.

---

[86] Russell and Copleston, 'Debate on the Existence of God', 175.
[87] Martin, *Thomas Aquinas*, 118.
[88] Nagel, 'Why Is There Anything?', 28.
[89] Martin, *Thomas Aquinas*, 146–53.

3. It is not possible that there should be an infinite series of efficient causes.
4. Therefore we must suppose that there is some prime efficient cause (*prima causa efficiens*), which is what everyone calls 'God'.

My concern here is not to evaluate the reliability of this argument[90]—which Aquinas clearly presents as a *ratio* for believing in God—but to point out that Aquinas sees it as offering a causal explanation of what is observed and experienced within the world. Although Aquinas's approach is ontic rather than epistemic, the context within which it is set presupposes some form of conceptual integration of both ontic and epistemic modes of explanation, even if this is not developed at this point in the *Summa Theologiae*.

This approach has, of course, been developed by many modern philosophers, who have argued for the existence of a transcendent necessary being that is the ground for an ultimate explanation of why particular contingent beings exist and undergo particular events.[91] Theism in general (and Christianity in particular) are held to consider it important 'to articulate a theoretical framework that makes possible ultimate explanation of reality'.[92] The question of how such a transcendent being relates to a specific notion of God—such as the 'God of the Christians' (Tertullian)—requires further discussion. Some have argued for developing a case for the existence of such a transcendent being on the basis of its capacity to explain the existence and character of the world, and then sought to correlate this with the Christian God; others have developed approaches which seek to demonstrate the explanatory capacity of the Christian understanding of God, seeing such a capacity as an indicator of such a God's existence. Yet this specific approach to explanation also includes a critical element, in that it poses the question of how a purely naturalist or materialist interpretation of our world can account for the appearance, through the operation of the laws of physics and chemistry, of conscious beings such as ourselves, who prove to be capable of discovering those laws and understanding the universe that they govern.[93]

---

[90] For a critical assessment, see Kenny, *The Five Ways*, 34–45.
[91] See especially O'Connor, *Theism and Ultimate Explanation*.
[92] O'Connor, *Theism and Ultimate Explanation*, 65.
[93] This is a recurring theme in Plantinga, *Where the Conflict Really Lies*.

THE IMAGE OF GOD AND RELIGIOUS EXPLANATION

We need to note a theological motif which has played a significant role in Christian reflection on the capacity of the human mind to interpret the world. It has often been noted that many leading scientists of the Renaissance saw theology as offering an imaginative template that enabled them to make sense of the world.[94] Yet many theologians regarded the notion of humanity bearing the 'image of God' as having important epistemic outcomes, including a propensity or capacity to discern God within creation. The idea of the 'image of God' is biblical;[95] yet the notion was developed in significant manners within the Christian theological tradition,[96] which often conceived it as a rational or imaginative template which enabled or facilitated the discernment of a theistic explanation of the world. In *Harmonices Mundi*, the theologically informed astronomer Johann Kepler argued that the fact of humanity's bearing the *imago Dei* predisposed humans to think mathematically, and thus grasp the structure of the created order:[97]

In that geometry is part of the divine mind from the origins of time, even from before the origins of time (for what is there in God that is not also from God?), it has provided God with the patterns for the creation of the world, and has been transferred to humanity with the image of God.

The 'image of God' is seen here not as an explanation in itself, but rather as a rational or imaginative template that facilitates such explanation in the first place. A similar view was taken by William Whewell, whose inductive scientific method reflects his belief that the 'Fundamental Ideas' which we use to organize our sciences resemble the ideas used by God in the creation of the physical universe. God

---

[94] See, for example, Hon, 'Kepler's Revolutionary Astronomy'.
[95] Middleton, *The Liberating Image*, 15–90.
[96] Waap, *Gottebenbildlichkeit und Identität*; Robinson, *Understanding the 'Imago Dei'*. For more recent interpretations, see Burdett, 'The Image of God and Human Uniqueness'.
[97] Kepler, *Gesammelte Werke*, vol. 6, 233.

Rational Explanation in Science and Religion 149

has created our minds such that they contain these ideas (or their 'germs'), so that 'they can and must agree with the world'.[98]

One of the most creative interpretations of the *imago Dei* was developed by Dorothy L. Sayers, who was convinced that Christianity gave her a tool by which she might 'make sense of the universe', disclosing hidden patterns and allowing meaning to be discerned within its otherwise opaque mysteries. Where *The Nine Tailors* (1934) addressed the 'mystery of the universe', *Gaudy Night* (1935) engaged the 'mystery of the human heart'.[99] Sayers's partly autobiographical work, *Cat O'Mary*, makes reference to this quest for meaning, and the intellectual pleasure it brought to its central character, Katherine. 'When Katherine sat down to prepare a passage of Molière, she experienced the actual physical satisfaction of plaiting and weaving together innumerable threads to make a pattern, a tapestry, a created beauty.'[100]

Sayers came to believe that such patterns were not human inventions, but represented an embedded pattern of the creative human mind,[101] itself echoing the deeper patterns of divine rationality. In *The Mind of the Maker* (1934), Sayers offered what is essentially an outworking of her own distinct notion of the 'image of God' in humanity as a kind of imaginative template, which predisposes human beings to think and imagine in certain ways.[102] The 'same pattern of the creative mind' is evident in both theology and art, and points to a deeper inbuilt imaginative template, which enables and encourages human beings to discern patterns in the deepest aspects of life. Sayers was inclined to think that the patterns of human creative processes 'correspond to the actual structure of the living universe', so that the 'pattern of the creative mind' is to be seen as an 'eternal Idea' rooted in the being of God.[103]

---

[98] Whewell, *On the Philosophy of Discovery*, 359.
[99] Kenney, *The Remarkable Case of Dorothy L. Sayers*, 53–119.
[100] Sayers, *Child and Woman of Her Time*, 97.
[101] Sayers, *The Mind of the Maker*, 171–3.
[102] Sayers, *The Mind of the Maker*, 15–24.
[103] Sayers, *The Mind of the Maker*, 172–3.

UNDERSTANDING AND EXPLAINING: A RELIGIOUS PERSPECTIVE

Earlier in this work (126–7), we briefly noted the complex (and contested) distinction between *Verstehen* and *Erklären*, broadly marking the boundaries of the social and natural sciences. So how does religious explanation fit into this dichotomy? It seems clear that most religious writers operate with a dual model of explanation, incorporating both these elements, affirming the importance of understanding and informing the complex world of lived experience from the point of view of those who actually inhabit it.

A Christian 'theory'—or, perhaps better, a Christian *imaginarium*[104]—thus aims to describe, explain, and enhance our understanding of the world, with a view to enabling and informing principled action within it. From what has already been said, such a theory has both ontic and epistemic elements; it also has existential outcomes, which are probably best framed in terms of the affirmation of meaning.

H. Richard Niebuhr's 1941 essay 'The Story of Our Lives' argued that Christians should focus on the capacity of the 'irreplaceable and untranslatable' narrative of faith to generate meaning.[105] This story was not an argument for the existence of God, but a simple recital of the events surrounding Jesus Christ, and an invitation to become part of that story. How participants understand being part of such a shared history is not framed in terms of a detached scientific explanation of life, but rather about a subjective, committed, and engaged attitude to existence, resting on a set of implicit assumptions that need to be unpacked and given systematic formulation. Niebuhr's essay saw a new interest emerge in the capacity of the Christian story to generate moral values and frameworks of meaning, which expanded (rather than marginalized) its ontic and epistemic dimensions.

The philosopher Iris Murdoch is one of many writers to emphasize the 'calming' and 'whole-making' effect of ways of looking at the

---

[104] For this term, see McGrath, *Re-Imagining Nature*, 41–7.
[105] Niebuhr, *The Meaning of Revelation*, 23–46.

world that, by generating a comprehensive vision of the world, allow it to be seen as ultimately rational and meaningful.[106] Christian theologians have, since the earliest times, argued that such seeming irrationalities as the presence of suffering in the world do not constitute a challenge to the notions of meaning and purpose that are embedded within the Christian faith. Augustine of Hippo, for example, set out an approach to the presence of evil within the world which affirmed the original integrity, goodness, and rationality of the world. Evil and suffering arose from a misuse of freedom, the effects of which are being remedied and transformed through redemption. Augustine argues that the believer is enabled to make sense of the enigmas of suffering and evil in the world by recalling its original goodness, and looking forward to its final renewal and restoration in heaven.

Yet the default position of contemporary Western culture tends to echo the view of the physicist Steven Weinberg, that the natural sciences disclose a meaningless universe. Scientific explanation entails the denial or evacuation of meaning. 'The more the universe seems comprehensible, the more it seems pointless.'[107] Nothing seems to fits together. There is no big picture. So do new scientific ideas destroy any idea of a meaningful reality? The English poet John Donne expressed similar anxieties in the early seventeenth century, as new scientific discoveries seemed to erode any sense of connectedness and continuity within the world.

The New Testament, however, speaks of all things 'holding together' or being 'knit together' in Christ (Colossians 1:17), thus suggesting that a hidden coherence lies beneath the external semblances of our world.[108] Christianity provides a framework which allows an affirmation of the *coherence of reality*. However fragmented our world of experience may appear, there is a half-glimpsed 'bigger picture' which holds things together, its threads connecting together in a web of meaning what might otherwise seem incoherent and pointless. This is a major theme in one of the finest Christian literary classics—Dante's *Divine Comedy*. As this great Renaissance poem

---

[106] Murdoch, *Metaphysics as a Guide to Morals*, 7.
[107] Weinberg, *The First Three Minutes*, 154.
[108] Tanzella-Nitti, 'La dimensione cristologica dell'intelligibilità del reale'.

draws to its close, Dante catches a glimpse of the unity of the cosmos, in which its aspects and levels are seen to converge into a single whole. This insight, of course, is tantalizingly denied to him from his perspective on earth; yet once grasped, this perspective enables him to see his work in a new light. There is a hidden web of meaning and connectedness behind the ephemeral and seemingly incoherent world that we experience.

This way of seeing things engages what is perhaps the greatest threat to any perception of meaningfulness in life or in our world—its seeming disorder and incoherence. Yet there is a deeper issue here. As we noted earlier in this work, John Dewey argued that the 'deepest problem of modern life' was our collective and individual failure to integrate our 'thoughts about the world' with our thoughts about 'value and purpose'.[109] If there is not an outright incoherence here, there is at least a *disconnection* between the realm of understanding the cognitive issue of how we and our world *function*, and the deeper existential question of what we and our world *mean*. Christianity offers a 'big picture' of reality which values and respects the natural sciences, while insisting that there is more that needs to be said about deeper questions of value and meaning. For Karl Popper, such 'ultimate questions' lay beyond the scope of the scientific method, yet were clearly seen as important by many human beings.

Perhaps the greatest threat to any sense of the coherence of reality is posed by the existence of pain and suffering. Christianity provides a series of possible mental maps which position illness and suffering in such a manner as to allow them to be seen as coherent, meaningful, and potentially positive, allowing them to foster personal growth and development. Some of these maps—such as those offered by Augustine of Hippo, Ignatius Loyola, and Edith Stein—portray illness as something that is not part of God's intention for humanity, but which can nevertheless be used as means of growth; other maps, such as that developed by Martin Luther, tend to hold suffering as something which God permits, with the objective of stripping away illusions of immortality and confronting human beings with the harsh reality of their frailty and transience.

---

[109] Dewey, *The Quest for Certainty*, 255.

Such Christian frameworks of meaning encourage a positive expectation on the part of believers that something may be learned and gained through illness and suffering. They make available new ways of thinking about life, and catalyse the emergence of more mature judgements and attitudes. Although this consideration has clear implications for Christian attitudes to illness and their outcomes, it is increasingly being recognized as being of significance in coping with ageing—an increasingly important phenomenon in Western culture.

For many of its leading theologians, Christianity is concerned with the meaningful inhabitation of our world, and offers a developed and nuanced understanding of what that meaning might be and how it plays out in real life. As we have argued in this closing section, it holds together ontic, epistemic, and existential aspects of the human quest to make sense of our world. To make sense of our world is not an end in itself, but the starting point for its meaningful inhabitation.

So what forms of reasoning allow us to make sense of our world? In the next chapter, we shall consider the scientific and theological application of deductive, inductive, and abductive modes of analysis.

6

From Observation to Theory

Deduction, Induction, and Abduction

The greatest stimulus to the exploration of the rationality of the universe is a sense of wonder at its immensity, beauty, strangeness, and solemnity.[1] Yet that sense of wonder proves generative, creating a desire to understand our beautiful and mysterious universe, and our own place within it. It precipitates a process of reflection, grounded in what we observe, stimulated by our sense of wonder, and directed towards grasping at least something of the greater vision of reality that lies behind what we can observe. There is a rich tradition of engagement with such questions within the Christian theological tradition, especially during the Middle Ages.[2] The beauty and ordering of nature were seen as outcomes of their divine creation, and reflecting the greater beauty and wisdom of God.[3] Indeed, many argue that this tradition of reflection did much to lay the ground for a religiously motivated emergence of the natural sciences in western Europe—for example, through the formulation of a natural theology which gradually transitioned into a natural philosophy.

So what rational strategies underlie a transition from the observation of our world to the identification and development of its

---

[1] Hesse, *'Mit dem Erstaunen fängt es an'*, 7–10; Falardeau, 'Le sens du merveilleux'; Evans, *Why Believe?*, 32.
[2] See, for example, Kretzmann, *The Metaphysics of Creation*; Hannam, *God's Philosophers*.
[3] McGrath, *Re-Imagining Nature*, 41–99.

patterns, and their potential implications?[4] The enterprise once known as 'natural philosophy'[5]—which has now been displaced by the somewhat different notion of 'natural science'—placed considerable emphasis on the discernment of intelligibility within nature, and the epistemic virtue of being able to identify the deeper structures which lay behind events and entities in the natural world:[6]

The hallmark of natural philosophy is its stress on intelligibility: it takes natural phenomena and tries to account for them in ways that not only hold together logically, but also rest on ideas and assumptions that seem right, that make sense; ideas that seem natural.

So how is this process of interpretation of our world to be carried out? One possibility is to use human reason to tell us—in effect to determine—what kind of world we inhabit. The natural sciences are to be seen as a principled rebuttal of such a procedure. Far from laying down in advance what the world ought to be like (on the basis of some presupposed philosophical or theological system), the natural sciences seek to discover what it is like by empirical investigation and rational reflection upon observations. The physical world is too complex to allow its structures, forms, or purposes to be predetermined by human reason—unless, of course, there is some compelling a priori reason for believing that the human mind can determine the structure of our world independently of any encounter or engagement with that world. No such compelling reasons have yet been adduced. The natural scientist—especially one who has struggled with quantum theory—is unlikely to ask 'Is it reasonable to believe this?' as if the human mind could determine beforehand the rational character of our world. The proper question is this: 'What evidence might lead me to believe this?'

---

[4] In practice, psychological research indicates that science does not use new or hitherto neglected modes of reasoning in exploring nature; it simply applies everyday modes of thinking in more precise ways to a specific set of observations. For further discussion, see Dunbar and Klahr, 'Scientific Thinking and Reasoning'; Dunbar, 'How Scientists Really Reason'; Feist, *The Psychology of Science and the Origins of the Scientific Mind*, 186–217.

[5] For the transition from 'natural philosophy' to 'natural science' in the nineteenth century, see Cahan, *From Natural Philosophy to the Sciences*.

[6] Dear, *The Intelligibility of Nature*, 173.

These considerations lay behind Karl Popper's emphatic declaration that *truth is above human authority*.[7] Popper was critical of the idea that we should justify our knowledge by 'positive reasons'—that is, by 'reasons capable of establishing them', which were deemed to be secure, and hence did not themselves require demonstration. Truth cannot be established by decree; only by investigation. Popper argued that recognizing that truth is above human authority was essential to good philosophy and good science. 'Without this idea there can be no objective standards of inquiry; no criticism of our conjectures; no groping for the unknown; no quest for knowledge.'[8] Popper's criticism of self-evident or foundational ideas mandates an empirical engagement with the natural world, demanding a rational engagement with observation, in order to develop theories. Yet neither reason nor observation are 'authorities'; they are tools to help us in the task of interpreting and understanding our world.[9] 'All theories are, and remain hypotheses: they are conjecture (*doxa*) as opposed to indubitable knowledge (*episteme*).'[10]

Popper's concerns need to be taken seriously, as we reflect on the process of moving from observation of our world to the development and evaluation of theories about that world. There is widespread concern within the scientific community that the way in which science is taught to teenagers in high schools tends to entail presenting science chiefly as *episteme* rather than *doxa*, as a body of reliable and stable information—thus failing to explain the corrigibility and provisionality of scientific theories, and the complexity of the journey from observation to theory.[11]

Science is often taught more as dogma—a set of unequivocal, uncontested and unquestioned facts—more akin to the way people are indoctrinated into a faith than into a critical, questioning community.

Both the manner in which observations are acquired and the way they are interpreted need critical examination. In this chapter, we shall

---

[7] Popper, 'Truth without Authority', 56–7.
[8] Popper, 'Truth without Authority', 57.
[9] Popper, 'Truth without Authority', 55.
[10] Popper, *Conjectures and Refutations*, 103–4.
[11] Osborne, 'Teaching Critical Thinking?', 54. See also Osborne, Simon, and Collins, 'Attitudes towards Science'.

reflect on the rational basis of theory development, and the character and reliability of the styles of reasoning that lie behind the formulation of theories.

Where Aristotle merely encouraged his readers to accumulate observations about the natural world, nineteenth-century empirical philosophers such as William Whewell pointed to the need to interpret and integrate them, thus discerning what patterns or greater picture might be intimated by those observations. In the early seventeenth century, Francis Bacon suggested that empirical philosophers—the word 'scientist' was not widely used until the late 1830s—are like ants, in that they simply accumulate observations. Rationalists, Bacon suggested, resemble 'spiders, who make cobwebs out of their own substance'. Yet for Bacon the true natural philosopher is like a bee, which 'gathers its material from the flowers of the garden and of the field, but transforms and digests it by a power of its own'.[12] For Bacon, observations had to be transformed and digested within the human mind. A process of analysis and synthesis underlay the transition from empirical observation to theoretical interpretation. Observation and theory were *connected*; they were nevertheless *distinct*.

In this chapter, we shall consider the process of theory development in the natural sciences and Christian theology, focussing especially on the rational criteria which guided and governed it. We begin by considering the way in which theories and observation are entangled.

THE ENTANGLEMENT OF THEORY AND OBSERVATION

It is widely agreed that there is 'intimate and inevitable entanglement' between observation and theory,[13] in that what is observed is modulated by both the human observer, and the theoretical link between an observed value and its interpreted significance. This point is

---

[12] Bacon, *The New Organon*, 79.
[13] Kuhn, 'Logic of Discovery or Psychology of Research?', 267.

important for two reasons: first, in assessing the difficulties inherent in attempting to represent our complex universe through the reducing lenses of scientific theories or models; and second, in reminding us how our observation of the universe is shaped by those theoretical lenses in the first place. The New Atheist philosopher A. C. Grayling argued that theological reasoning was unacceptable to a rational person, because it was undertaken within 'the premises and parameters' of a system.[14] Yet this overlooks the obvious fact that most human thinking—particularly scientific discourse—is elaborated within 'the premises and parameters' of some system or other, including mathematics and logic. Observation and interpretation intertwine in an inescapable circularity.

For example, the current estimate of the age of the universe is roughly 13.8 billion years. But how do we know this, in the absence of any continuous chronological monitoring of this history? After all, in 1919, the universe was thought to be of indefinite or infinite age; in 1929, based on an early determination of Hubble's constant, it was believed to be two billion years old. Current estimates of the age of the universe are based on observations that are interpreted within 'the premises and parameters' of the Lambda-CDM model.[15] The observations themselves tell us nothing about the age of the universe.

Science does not read off the age of the universe directly; rather, it interprets certain observations within the framework of the Lambda-CDM (or 'concordance') model to derive the age of the universe. The speeds and distances of those galaxies are also not observed directly, but are inferred on the basis of 'the premises and parameters' of additional physical theories—such as the correlation between velocity and the Doppler red shift.[16] Yet the instrumental observation of certain parameters presupposes some (provisional and corrigible) theoretical correlation between an observed parameter and a second parameter which is inaccessible to observation.[17]

---

[14] Grayling, *The God Argument*, 66.

[15] This model incorporates certain important assumptions, most notably that the universe is homogenous and isotropic. For the problems that this raises, see Merritt, 'Cosmology and Convention'.

[16] Harrison, 'The Redshift-Distance and Velocity-Distance Laws'.

[17] For the philosophical issues this raises, see van Fraassen, *Scientific Representation*, 93–111.

Furthermore, the Lambda-CDM model faces conceptual difficulties, most notably a long-standing concern about significant anomalies, such as the following:[18] the 'Lithium Problem', the 'Core–Cusp Problem', the 'Missing Satellites Problem', the 'Missing Baryons Problem', and the 'Too Big to Fail Problem'. Although these are generally presented as problems that remain to be solved from within the existing Lambda-CDM model, it is possible to argue that they really ought to be regarded as falsifications of the model.[19]

This entanglement of theory and observation has potentially significant implications for the generation and testing of theories. If observation is 'theory-laden', how can we speak of 'observationally equivalent theories'?[20] If scientific observations depend on presumed theoretical frameworks, an objective empirical test of theories and hypotheses by independent observation and experience would seem to be impossible owing to the implicit circularity of any possible empirical assessment of the theory. In practice, once this difficulty is realized, and its negative implications for a naïve scientific positivism are appreciated, strategies can be developed to deal with it.[21]

LOGICS OF DISCOVERY AND JUSTIFICATION

In an influential essay of 1973, Mary Hesse suggested that natural science was to be thought of as essentially a learning device consisting of a receptor, a theorizer, and a predictor.[22] Empirical information is received from the environment, and stimulates a process of analysis and synthesis, leading to the generation of theories, and subsequently to evaluating such theories through considering how its account of reality correlates with experience. The second two elements of Hesse's analogy correspond to the processes generally described as the logics of discovery and justification.

---

[18] Merritt, 'Cosmology and Convention', 42.
[19] See the careful argument presented in Kroupa, 'The Dark Matter Crisis'.
[20] Adam, *Theoriebeladenheit und Objektivität*, 25–50.
[21] See, for example, the constructive proposals set out in Buzzoni, 'Erkenntnistheoretische und ontologische Probleme der theoretischen Begriffe'.
[22] Hesse, 'Models of Theory Change'.

The natural sciences represent a complex intellectual enterprise that essentially consists of two interdependent episodes: the first, developing a hypothesis, is essentially imaginative or creative; the second, in which such a hypothesis is interrogated, is fundamentally critical. To advance a hypothesis is a creative exercise. However, such scientific hypotheses or theories must also be subject to critical examination and empirical testing.

A 'theory' can be understood as an imaginative conjecture of what might be true that provides the intellectual incentive to investigate whether this might indeed be the case. Karl Popper's brief account of the nature of theory serves as an excellent starting point for our discussion.[23]

> Scientific theories are universal statements. Like all linguistic representations they are systems of signs or symbols... Theories are nets cast to catch what we call "the world"; to rationalize, to explain and to master it. We endeavor to make the mesh ever finer and finer.

Popper's approach is, of course, open to further development—for example, we should be noting how theory can used as a 'sensitizing device' to view and inhabit the world in a certain way, thus leading to a cultivation of attentiveness towards it.[24] Yet for the purposes of this chapter, Popper's definition is perfectly serviceable. It allows us to ask this question: what intellectual processes and conventions lead from an observation of the world to the acceptance of such 'universal statements'? In this chapter, we shall consider the three types of reasoning—deductive, inductive, and abductive—which might be involved in this process, and indicate their respective strengths and weaknesses, both in the natural sciences and their counterparts in theology.

We must, however, first give some thought to the process by which possible theoretical models might be developed. The great French biologist Claude Bernard (1813–78) identified the two stages of the scientific method as the formulation of a testable hypothesis and the subsequent testing of such a hypothesis by observation and

---

[23] Popper, *The Logic of Scientific Discovery*, 37–8.
[24] Klein and Myers, 'A Set of Principles for Conducting and Evaluating Interpretive Field Studies in Information Systems', 75.

experiment. Bernard was clear that such scientific theories remained partial and provisional. Theories, he declared, were 'only partial and provisional truths' which represent nothing more than 'the current state of our understanding and are bound to be modified by the growth of science'.[25]

A historical example nicely illustrates the questions that need to be raised. The structure of benzene (C_6H_6) puzzled nineteenth-century chemists. How could its highly polyunsaturated structure be represented theoretically? In 1865, August Kekulé set out the radical new idea that benzene possessed a cyclical structure. Kekulé did not explain the 'logic of discovery' lying behind this idea at that time, although his subsequent work provided an extensive 'logic of justification' for the ring structure of benzene. It was only in 1890, at a celebration marking the twenty-fifth anniversary of this suggestion—by then widely accepted—that Kekulé explained how this idea came to him.[26] While drowsing in front of his fire, he had a vision of a snake chasing its own tail, which suggested an annular structure. While the origins of this idea might indeed be somewhat speculative, even mystical, the fact remains that, when such a structure was proposed for benzene and was checked out against the evidence, it seemed to account for it satisfactorily. The manner of its derivation might be opaque; the manner of its verification was, however, clear—and ultimately persuasive.

Many have found Kekulé's dream helpful in trying to understand the process by which new ways of understanding our world or envisaging reality arise.[27] Kekulé's new way of envisaging the molecular structure of benzene was unquestionably grounded in a deep knowledge of its chemical properties; yet what crystallized and integrated that knowledge was an act of imaginative discernment, rather than rational dissection. Considerations such as this have led some to conclude that it is not possible to prescribe a single logical method or research practice that produces new ideas, or to reconstruct logically

---

[25] Bernard, *Introduction à l'étude de la médecine expérimentale*, 64–80.
[26] Kekulé, 'Benzolfest Rede'. 'Eine der Schlangen erfasste den eigenen Schwanz und höhnisch wirbelte das Gebilde vor meinen Augen.'
[27] See, for example, Fischer, *Wie der Mensch seine Welt neu erschaffen hat*, 190–2.

the process of discovery.[28] The importance of the imagination in scientific discovery has long been recognized, despite attempts to marginalize or suppress this in the interests of a rationalist agenda.[29] 'The act of discovery escapes logical analysis; there are no logical rules in terms of which a "discovery machine" could be constructed that would take over the creative function of the genius.'[30] This suggests that the process of articulating and developing new theories is best seen, not as a philosophical, but as a psychological question.

However, such imaginative leaps and correlations represent only one of several approaches to the development of theories, and they must not be considered either typical or normative. Indeed, despite the obvious examples of scientific discovery that are clearly accidental or owe their origins to an act of imagination, some philosophers of science would still argue that hypothesis generation and theory construction are fundamentally and characteristically rational processes.[31]

Once a theory is proposed, it needs to be evaluated. At this point, a 'logic of justification' enters the picture, in that the proposed new theory can be subjected to a number of tests, such as its internal coherence, its capacity to represent observations, and its ability to make novel predictions or generate research programmes. Some see this process as a rule-based process, where others see the process of justification as implicitly requiring elements of discovery. For example, in order to justify a theory, it is necessary to 'discover' an appropriate test or set of criteria adapted to this new theory.[32]

There has been much discussion about the criteria that might be used to justify a theory. Irritated by what he regarded as the excessive plasticity of Freudianism and Marxism, Karl Popper proposed falsifiability as a criterion of demarcation between the empirical sciences and other forms of knowledge.[33] Popper's argument that scientific hypotheses should be falsifiable rather than simply verifiable posits an

---

[28] Popper, *The Logic of Scientific Discovery*, 7–8.
[29] See especially the classic account of Beveridge, *The Art of Scientific Investigation*, 53–63.
[30] Reichenbach, *The Rise of Scientific Philosophy*, 231.
[31] See, for example, Nersessian, 'How Do Scientists Think?'
[32] As argued by Nickles, 'Discovery'.
[33] Díez, 'Falsificationism and the Structure of Theories'.

asymmetry between the falsifiability and the verifiability of universal statements: a universal statement can be falsified if it is found to be inconsistent with a single observation; such a statement cannot, however, be proven to be true by virtue of the truth of an array of particular observations, no matter how numerous these may be.

Yet falsifiability is perhaps a blunt tool; as Pierre Duhem pointed out, a physicist simply is not in a position to submit any given isolated hypothesis to experimental test, in that the experiment can only indicate that one hypothesis *within a larger group of hypotheses* requires revision—but not which specific element of this ensemble of hypotheses is at fault. 'An experiment in physics can never condemn an isolated hypothesis but only a whole theoretical group (*un ensemble théoretique*).'[34] Quine's reformulation of Duhem's argument makes the same point succinctly: 'our statements about the external world face the tribunal of sense experience not individually, but only as a corporate body'.[35]

A historical example which illustrates the difficulties associated with falsification is provided by the observed orbital perturbations of the planet Uranus. The observed motion of Uranus, discovered in March 1781 (although observed earlier, and mistaken for a star), was not what was predicted on the basis of Newtonian mechanics. Popper thus concluded that this was a prima facie case of the falsification of Newton's gravitational theory.[36] Others at the time, however, held that a possible explanation lay in challenging the uninterrogated assumption that no planets lay beyond Uranus, and that its observed orbital perturbation might reflect the gravitational influence of a trans-Uranic planet. This possibility was championed by Alexis Bouvard (1767–1843); calculations of mathematicians in England and France led to the discovery of the trans-Uranic planet Neptune in 1846. Once more, the planet had been observed earlier, but had again been mistaken for a fixed star. For Popper, the hypothesis of a trans-Uranic planet represented an evasion of the falsification of Newton's theory of gravitation; for others, of course, it represented

---

[34] Duhem, *La théorie physique*, 284.
[35] Quine, 'Two Dogmas of Empiricism', 41.
[36] For an excellent analysis, see Bamford, 'Popper and His Commentators on the Discovery of Neptune'.

an attempt to find what aspect of Newton's theory needed revision, rather than its wholesale rejection. This second view has prevailed as a more reliable account of scientific practice.

The relation of the processes of discovery and justification is agreed to be complex,[37] especially when the contexts of both discovery and justification are taken into account. Yet although it is clearly helpful to distinguish them, they cannot entirely be separated. In essence, a distinction is drawn between the processes of *conceiving* a theory and *validating* that theory—in other words, establishing its epistemic support.

Natural scientists use a variety of reasoning processes in developing and validating theories. Three of these may be singled out for special consideration: deductive, inductive, and abductive patterns of reasoning. Charles Peirce suggested that their relationship in the logic of scientific discovery might be conceived in the following way: discovery begins with abduction, in which some hypothesis is formulated to explain some problem; next comes deduction, in which this hypothesis is rendered precise and predictions are deduced; and finally induction, in which the hypothesis is tested by experience.[38] Yet although these three processes of thought are typical of the sciences, they are only distinctively scientific in relation to their objects of investigation, and are regularly used in other contexts. We shall consider each in what follows.

DEDUCTION IN THE NATURAL SCIENCES

The natural sciences deal with an accumulation of observations, leading to reflection on what patterns of correlation or association might be identified, and hence how they might be explained—an issue which we considered in some detail in the previous chapter. This suggests that the most natural reasoning processes that are

---

[37] See especially Schickore and Steinle, eds, *Revisiting Discovery and Justification*; Hoyningen-Huene, 'Context of Discovery and Context of Justification'.
[38] Rodrigues, 'The Method of Scientific Discovery in Peirce's Philosophy'.

appropriate to the natural sciences are inductive and abductive, rather than the forms of deductive argument that are based on general principles.

Furthermore, the syllogistic structure of deductive reasoning entails the risk of at least one of its premises, once thought to be self-evidently true, being shown to be corrigible and provisional, subject to correction or even repudiation through more sophisticated methods of investigation or a general increase in the available evidence. A trivial familiar example illustrates this point:

Premise 1: All swans are white.
Premise 2: I observe a swan.
Conclusion: The swan I observe is white.

The deductive force of this might have seemed self-evident to Europeans in the seventeenth century. However, reports of the discovery of black swans in Western Australia in 1697 by Dutch explorers led by Willem de Vlamingh called into question the empirical basis of the first of the premises,[39] and the reliability of the conclusion. The swan in question was *probably* white, on statistical grounds; it was not, as once might have been thought, *necessarily* white. The problem, as Hume pointed out, is that no finite number of empirical observations can be held to entail an absolute or secure generalization.[40]

Yet while there is some truth in this general observation, it is important to note that deductive forms of reasoning play a critical role in the Deductive-Nomological approach to scientific explanation (136). This approach, which was highly influential in the 1950s and 1960s,[41] is especially associated with Carl G. Hempel (1905–97). For Hempel, empirical phenomena could be explained by demonstrating that they fit into a 'nomic nexus of observed regularities'.[42] For Hempel, scientific explanation 'aims at showing that the event in

---

[39] Popper, *Logic of Scientific Discovery*, 82–3. A similar point was made by John Stuart Mill in his *System of Logic* (1843).
[40] Boulter, *The Rediscovery of Common Sense Philosophy*, 77.
[41] For comment on its influence, see Salmon, *Four Decades of Scientific Explanation*, 3–89.
[42] Hempel, *Aspects of Scientific Explanation*, 488.

question was not a "matter of chance", but was to be expected in view of certain antecedent or simultaneous conditions'.[43]

For Hempel, the role of laws in deductive-nomological explanation is to connect the *explanandum* with the particular *explanans*. Not all explanations are causal; the key point is to establish a 'nomic nexus' which explains what is observed. An explanation is thus essentially a deductive derivation of the occurrence of the event to be explained from a set of true propositions, which include at least one statement of a general scientific law. Such an explanation 'shows that, given the particular circumstances and the laws in question, the occurrence of the phenomenon was to be expected; and it is in this sense that the explanation enables us to understand why the phenomenon occurred'.[44]

There are, of course, some concerns about this approach. Consider, for example, Hempel's contrasting of two statements:[45]

1. All members of the Greensbury School Board for 1964 are bald.
2. All gases expand when heated under constant pressure.

Hempel argues that if the first such statement is true, it is so accidentally, and cannot be used as the basis of a deductive argument. The second of these generalizations, however, has the status of a *natural law*, and thus has explanatory capacity. The *explanans* can account for the *explanandum* in the second case, but not the first. Yet what if it were unexpectedly shown that there were exceptions to the apparent generality that gases expand when heated under constant pressure? It is, of course, improbable; yet the history of the natural sciences is replete with once seemingly secure generalizations which were subsequently shown to have exceptions (such as black swans)—exceptions which often served as gateways to a deeper understanding of the phenomena in question.

Hempel's approach is now generally seen as of historical interest, representing an early phase in scientific explanation which has given way to more reliable alternatives which place greater emphasis on

---

[43] Hempel, *Aspects of Scientific Explanation*, 235.
[44] Hempel, *Aspects of Scientific Explanation*, 337.
[45] Hempel, *Aspects of Scientific Explanation*, 339.

causal approaches to explanation.⁴⁶ The question is whether the *explanans* which Hempel sees as providing a *deductive* basis for explanation is actually the outcome of an incomplete *inductive* sequence, which is necessarily open to correction and supplementation over time, as a result of the accumulation of observations. Inductive reasoning fails to arrive at universal truths, which can become the basis of secure processes of deduction. No matter how many singular statements may be accumulated, no universal statement can be logically justified on the basis of such an accumulation of observations. Even if all swans hitherto observed are white, it remains a logical possibility that the next swan will not be white.

Yet the importance of deductive reasoning within the natural sciences is reaffirmed by the so-called 'Hypothetico–Deductive Method', which in effect proposes a hypothesis for testing on the basis of the assumption that it is correct, and proceeds to enquire what observations might be expected to follow if it is indeed true. Such a hypothesis might be empirically falsified by determining whether or not certain logical consequences of the hypothesis, ascertained by deduction, are found to agree with the state of affairs actually found in the empirical world.⁴⁷ Yet even in this rational strategy, some form of induction seems to be entailed in the generation of legitimate hypotheses in many cases.

DEDUCTION IN CHRISTIAN THEOLOGY

While the rationality of Christian theology can be explored in many ways, attention has traditionally focussed on the specific issue of the rationality of theistic belief. While many still speak of 'proofs for the existence of God', it seems clear that Anselm of Canterbury's 'ontological argument' and Thomas Aquinas's 'Five Ways' and related approaches are really affirmations of the rationality of belief in God.⁴⁸

---

⁴⁶ Akeroyd, 'Mechanistic Explanation Versus Deductive-Nomological Explanation'.
⁴⁷ For evaluations of this approach, see Gemes, 'Hypothetico-Deductivism'; Sprenger, 'Hypothetico-Deductive Confirmation'.
⁴⁸ Plantinga, *The Nature of Necessity*, 221: 'Our verdict on these reformulated versions of St. Anselm's argument must be as follows. They cannot, perhaps, be said

Religious apologists of any opinion—including Christian apologists—are drawn to deductive arguments for the existence of God, in that such a priori arguments do not depend upon external validation for support. The validity of a deductive argument can be ascertained before, or even without, empirical validation, in that it rests on statements that are true in themselves. Many regard Anselm of Canterbury's so-called 'ontological argument'—although it must be stressed that Anselm did not describe it in such terms—as having the status of a deductive argument based on a priori truths. The argument may be set out as follows:[49]

1. God is the greatest possible being—something than which nothing greater can be conceived.
2. If God exists only in the mind as an idea, then a greater being could be imagined to exist both in the mind and in reality.
3. This being would then be greater than God.
4. God therefore cannot exist only as an idea in the mind.
5. God therefore exists both as an idea and in reality.

The argument is open to criticism—indeed, some would argue, to logical inversion. Yet it retains a certain fascination, in that it articulates an idea which has obstinately refused to go away.

Yet there are other deductive approaches which merit consideration. Classic versions of the cosmological argument, as set out by Gottfried Wilhelm Leibniz (1646–1716), have appealed to the 'Principle of Sufficient Reason' in developing a deductive argument that is held to succeed in demonstrating the existence of God. Other deductive arguments from contingency have aimed to establish the existence of a necessary or non-contingent being as an explanation of the existence of contingent beings.[50] Of particular interest are recent

---

to *prove* or *establish* their conclusion. But since it is rational to accept their central premise, they do show that it is rational to accept that conclusion.'

[49] For an excellent summary of Anselm's argument and its interpretation and assessment, see Leftow, 'The Ontological Argument'.

[50] Gale and Pruss, 'A New Cosmological Argument'.

developments of the *kalām* cosmological argument which are widely regarded as falling into the category of deductive thinking. Although such arguments originated from within an Islamic context, they have recently been reframed by the Christian philosopher William Lane Craig.[51] In general terms the argument can be set out as follows:[52]

1. Whatever begins to exist has a cause of its existence.
2. The universe began to exist.
3. Therefore, there exists a cause of the existence of the universe.

Craig then argues that since no scientific explanation (framed in terms of physical laws) can provide a causal account of the origin of the universe, the cause must be personal. In other words, an explanation is to be given in terms of a personal agent—which is clearly indicative of some form of theism, such as Christianity. Craig's particular way of formulating this argument has been criticized for some of its assumptions, such as a dynamic theory of time and the metaphysical impossibility of an actual infinite; it can, however, be reframed to avoid these particular concerns.[53]

Premise (1) is stated as if it were an unquestionable a priori truth, which can lead to certain reliable deductions. The idea that everything is caused by something is, of course, a widely accepted notion, and underlies ontic approaches to scientific explanation, noted earlier (129). In a letter of February 1754, possibly written to John Stewart, then Professor of Natural Philosophy at Edinburgh University, David Hume declared that he 'never asserted so absurd a Proposition as that any thing might arise without a Cause',[54] clearly regarding this as unthinkable. Yet it is unclear how Hume defended this position. Hume asserts that his certainty of the falsity of the proposition that anything might arise without a cause arises 'neither from Intuition nor Demonstration, but from some other Source'. Although at times Craig himself appears to treat this principle as an empirical

---

[51] Craig, *The Kalām Cosmological Argument*. See also Copan and Craig, eds, *The Kalām Cosmological Argument*.
[52] For a variant formulation, see Loke, *God and Ultimate Origins*, 85–107.
[53] Oderberg, 'Traversal of the Infinite, the "Big Bang" and the Kalām Cosmological Argument'.
[54] Hume, *Letters*, vol. 1, 187. The original manuscript of this letter does not include the name of the addressee: see Hume, *Letters*, vol. 1, 185, note 5.

generalization based on our ordinary and scientific experiences, or perhaps as an intuition that something cannot come out of nothing. Yet this involves a subtle change in the epistemic status of Premise (1) from an a priori truth to an a posteriori generalization.

Premise (2) is presented as a factual statement; it is, in fact, the *regnant interpretation* of a suite of observations of the universe, which achieved dominance in the second half of the twentieth century. It is not a self-evident truth, and would not have been accepted as true by the scientific community during earlier periods of history.[55] The idea that 'the universe began to exist' is not itself an observation, but an interpretation of observations, and once more lacks the status of being an a priori truth capable of bearing substantial epistemic weight.

Yet although such arguments may lack the certainty that some hope for, they certainly represent forceful affirmations of the rationality of theistic belief, even if there remains a 'gap' between the philosophical notion of an eternal necessary being and the specifically Christian conception of God.

INDUCTION IN THE NATURAL SCIENCES

For many philosophers of science, inductive reasoning is both the hallmark and the point of vulnerability of scientific thinking. In essence, inductive reasoning involves the observation of a series of events and attempts to discover what theoretical framework might best accommodate them. Such a theoretical framework, once inferred, can serve to explain future observations—just as anomalous observation may serve to disconfirm the original conclusions.

The principle of inductive thinking is easily stated, yet proves more difficult to apply. Many writers have noted the lack of clarity concerning both the process by which such extrapolation occurs, the intellectual mechanisms by which theoretical frameworks might be derived, and the criteria to be used in developing such frameworks.[56]

---

[55] Kragh, *Conceptions of Cosmos from Myths to the Accelerating Universe*, 125–63.
[56] Nickerson, 'Teaching Reasoning', 410–11.

From Observation to Theory

Widely accepted standards are available for evaluating the quality of deductive reasoning; however to evaluate inductive reasoning is a matter of considerable debate. Whitehead's reference to the theory of induction as 'the despair of philosophy' reflects the frustration that scholars have experienced in trying to codify inductive reasoning; but recognition of the importance of this type of reasoning is seen in Polya's observation that 'in dealing with problems of any kind, we need inductive reasoning of some kind'.

This palpable sense of frustration arises from a recognition of David Hume's critique of inductive methods,[57] set alongside the brute fact that, despite its manifest and manifold intellectual shortcomings, the method seems to deliver results. This dilemma is nicely captured by Charles Darwin, whose *Origin of Species* (1859) can be interpreted as one of the nineteenth century's most successful applications of inductive reasoning. Darwin was alert to the criticisms offered of this form of reasoning, but held that its pragmatic value could hardly be overlooked. 'It has recently been objected that this is an unsafe method of arguing; but it is a method used in judging the common events of life, and has often been used by the greatest natural philosophers.'[58]

While Darwin may have overstated his use of Baconian inductivism to enhance the public credibility of his argument,[59] there is little doubt that Darwin's approach is fundamentally inductive at many points. For Darwin, the test for his theory of natural selection was its capacity to accommodate a large range of biological phenomena, including some which were otherwise obscure or puzzling, when interpreted in terms of rival theories, such as special creation or Lamarckian transformism. 'Light has been thrown on several facts, which on the belief of independent acts of creation are utterly obscure.'[60]

So what observations did Darwin have in mind? *The Origin of Species* set out a series of biological phenomena which his theory of natural selection seemed to explain in a more elegant and less forced manner than its two main alternatives. How, for example, was the

---

[57] Millican, 'Hume's Sceptical Doubts Concerning Induction'.
[58] Darwin, *Origin of Species*, 444.
[59] As argued by Ayala, 'Darwin and the Scientific Method'.
[60] Darwin, *On the Origin of Species by Means of Natural Selection*, 164.

uneven geographical distribution of life forms throughout the world, especially marked in the peculiarities of island populations, to be accounted for? Or how could the persistence of 'rudimentary structures', which have no apparent or predictable function—such as the nipples of male mammals, the rudiments of a pelvis and hind limbs in snakes, and wings on many flightless birds—be explained? Darwin's theory offered an explanatory elegance which he considered to be superior to rival interpretations of these observations. Darwin's argument was helped to no small extent by the plausible analogy between the familiar breeding technique of 'artificial selection' and the hypothesized mechanism of 'natural selection'.[61]

Detailed studies of other scientists' research practices and rational strategies have confirmed the role of inductive thinking within the natural sciences. Paul Thagard, for example, studied how such a form of reasoning led Robin Warren and Barry Marshall to propose that peptic ulcers are generally caused, not by excess acidity or stress, but by a bacterial infection.[62] Thagard, however, was careful to note how this research involved research strategies that combined serendipity, questioning, and search.

Induction, by its nature, can never deliver certainty. Hume's critique of inductivism remains important—yet most natural scientists argue that pragmatically it makes little difference to the scientific method. But what of induction in theology?

INDUCTION IN CHRISTIAN THEOLOGY

Inductive approaches to demonstrating the rationality of Christian belief can be found throughout history, although these gained prominence in the aftermath of the scientific revolution. The apologetic strategy often known as 'physico-theology', which became increasingly important in the early eighteenth century, argued inductively from the perceived order or beauty of the natural world to the

---

[61] Largent, 'Darwin's Analogy between Artificial and Natural Selection in the *Origin of Species*'.
[62] Thagard, *How Scientists Explain Disease*, 39–97.

inference that a God lay behind both its existence and its character.[63] Although such approaches often inferred divine origination from the structures of the physical world, some developed such arguments based on an appeal to the beauty and complexity of the biological realm. The most important and influential of these, however, was developed in the first decade of the nineteenth century.

William Paley's *Natural Theology* (1802) is essentially an inductive argument for the existence of God as 'contriver'—that is, designer and constructor—of the natural world, based on an accumulation of observational data and reflection. The fundamental argumentative strategy is that the best explanation of such complex structures is a creator God.[64] While the work gained popular traction through its skilful deployment of the analogy of God as a watchmaker, its core logical argument is that nature shows such evidence of 'contrivance' that its complexity and functionality cannot be ascribed to chance.[65] Darwin's subsequent suggestion that such complex and adapted structures might evolve through natural means seemed to many to evacuate Paley's argument of its imaginative plausibility.[66]

In recent years, Christian philosophers and apologists have become increasingly interested in inductive approaches to the rationality of faith, especially in relation to the question of the existence of God, while avoiding the scientific pitfalls associated with earlier attempts. Richard Swinburne, for example, develops an essentially inductive approach to the rationality of faith, arguing that the existence of God is to be seen as the 'best explanation' of what is observed within the world, when seen as part of a larger cumulative case. For Swinburne, the existence of the universe can be made comprehensible if we suppose that it is brought about by God.[67]

Swinburne sets this inductive approach within a wider framework, grounded in the core belief that 'the fact that there is a universe needs explaining'. So which explanation is the best? Since 'it is reasonable to suppose' that there is an explanation of the universe in the first

---

[63] Harrison, 'Physico-Theology and the Mixed Sciences'; Mandelbrote, 'The Uses of Natural Theology in Seventeenth-Century England'.
[64] Gliboff, 'Paley's Design Argument as an Inference to the Best Explanation'.
[65] McGrath, 'Chance and Providence in the Thought of William Paley'.
[66] Ayala, 'In William Paley's Shadow'.
[67] Swinburne, *The Existence of God*, 9–10.

place,[68] Swinburne suggests that the question to be determined is which of two rival theories is more reasonable: the naturalist view that science can provide a natural explanation for the existence of this universe, or the theistic view that the universe and its phenomena exist because of the intentional causal activity of a personal being.

Swinburne's task is thus to identify possible explanations of the universe, and to determine which of these is 'best'. In making this decision, Swinburne does not see himself as required to prove the existence of God; rather, simply to show that the existence of God, however unlikely this might appear as an independent hypothesis, is better at explaining our nexus of observations and experiences. A priori, theism might perhaps seem very unlikely; yet, Swinburne argues, it is far more likely than its explanatory rivals. In developing this kind of inductive cosmological argument, however, Swinburne makes the (contested) criterion of simplicity the deciding factor between competing hypotheses concerning the existence of the universe.[69]

One of Swinburne's most significant contributions to assessing the rationality of theistic belief is his articulation of the case for theism in terms of Bayes's theorem. Swinburne holds that the evidence considered raises the overall probability of the theistic hypothesis above 0.5. Yet the deployment of Bayes's theorem raises some questions that are not entirely resolved in Swinburne's discussion. The use of Bayesian approaches requires, for example, agreement concerning the prior probability of the thesis under assessment and the probability of each piece of evidence given as background knowledge. There is no consensus on the prior probability of there being a God, which remains a topic of debate and discussion.[70] Nevertheless, Swinburne's overall project is unquestionably one of the most important exhibitions of a rational defence of theism to appear in recent years, highlighting how a range of inductive tools can be deployed in affirming the fundamental rationality of theistic belief.

---

[68] Swinburne, *The Existence of God*, 75.

[69] Ostrowick, 'Is Theism a Simple, and Hence, Probable, Explanation for the Universe?'

[70] Some of these issues are considered in an excellent collection of essays edited by Swinburne: see Swinburne, ed., *Bayes's Theorem*. Especially in his *Warranted Christian Belief* (2000), Alvin Plantinga is critical of Bayesian approaches: see Portugal, 'Plantinga and the Bayesian Justification of Beliefs'.

ABDUCTION IN THE NATURAL SCIENCES

The third reasoning method which plays an important role in the natural sciences has come to be known as 'abduction'. According to the philosopher and scientist Charles S. Peirce (1839–1914), abduction is the process by which observational activity leads to a process of imaginative generation, with the object of determining what intellectual frameworks might make sense of them. Peirce's model of discovery takes the individual as a cognizing agent in the world, whose knowledge arises through sensory experience, and is developed through model building and empirical testing.[71] As Peirce understands this approach, abduction is basically a kind of creative 'search strategy' which leads to the identification or creation of some 'promising explanatory conjecture which is then subject to further test'.[72] In an unpublished and undated lecture, he described abduction as[73]

... that process in which the mind goes over all the facts the [sic] case, absorbs them, digests them, sleeps over them, assimilates them, dreams of them, and finally is prompted to deliver them in a form, which, if it adds something to them, does so only because the addition serves to render intelligible what without it, is unintelligible.

For Peirce, abduction is a quest for a plausible framework for the accommodation of surprising observations. 'By plausible I mean that a theory that has not yet been subjected to any test, although more or less surprising phenomena have occurred which it would explain if it were true, is in itself of such a character as to recommend it for further examination.'[74] The characteristic trajectory of such an abductive approach could thus be set out as follows.[75]

---

[71] Kirklik and Storkerson, 'Naturalizing Peirce's Semiotics', 34–5.
[72] Schurz, 'Patterns of Abduction', 205. More generally, see Paavola, 'Abduction as a Logic of Discovery'; Magnani, *Abduction, Reason, and Science*; McKaughan, 'From Ugly Duckling to Swan'.
[73] MS 857: 4–5; cited in McKaughan, 'From Ugly Duckling to Swan', 466. On the creative aspects of abduction, see Prendinger and Ishizuka, 'A Creative Abduction Approach to Scientific and Knowledge Discovery'.
[74] Peirce, *Collected Papers*, vol. 2, 662.
[75] Peirce, *Collected Papers*, vol. 5, 189.

1. The puzzling phenomena A and B are observed.
2. But if C were true, then A and B would follow as a matter of course.
3. Hence, there is reason to suspect that C is true.

This naturally raises the question of how abduction functions as a 'logic of discovery'. How is C to be generated? How might a given epistemic framework be developed and proposed that might accommodate A and B, before subsequently being subjected to a process of critical evaluation? Peirce himself is quite clear that there are multiple possibilities here—including those which he categorizes in terms of inspiration and imagination. There are multiple logics of discovery; nevertheless, they are all to be subjected to a logic of verification, in which any proposed framework of interpretation is checked out against the observable facts. Theories are not to be judged by how they are devised, but by their capacity to accommodate observation and experience.

A related approach is found in N. R. Hanson's reflections on the advance of scientific knowledge, which identified three common elements within 'the logic of scientific discovery':[76]

1. The observation of some 'surprising' or 'astonishing phenomena', which represent anomalies within existing ways of thinking. This 'astonishment' may arise because the observations are in conflict with existing theoretical accounts.
2. The realization that these phenomena would no longer seem to be astonishing if a certain hypothesis (or set of hypotheses) H pertained. These observations would be expected on the basis of H, which would act as an explanation for them.
3. The conclusion that there is therefore good reason to for proposing that H be considered to be correct, and proceeding to confirm this by appropriate means.

Like Peirce, Hanson identifies astonishing or surprising observations as a fundamental motivation in stimulating and guiding the

---

[76] Hanson, 'Is There a Logic of Scientific Discovery?', 104. See also Schaffner, *Discovery and Explanation in Biology and Medicine*, 11–13. For the abductive aspect of Hanson's approach, see Paavola, 'Hansonian and Harmanian Abduction as Models of Discovery'.

enterprise of scientific discovery. Is there a theoretical standpoint from which these observations would not be astonishing, or even merely anomalous, but would be *expected*?

For Peirce, abduction is a creative and corrigible process, representing the 'provisional adoption of an explanatory hypothesis' as a way of making sense of a set of observations.[77] Abduction is a distinct and *generative* form of logical inference, which is the 'only kind of argument which starts a new idea'[78] or serves as 'the process of forming explanatory hypotheses'.[79] It often consists of an 'act of insight' that 'comes to us like a flash'.[80] Peirce's language here suggests that abduction can be compared to the creative and aesthetic dimensions of human perception, in which the explanatory hypothesis 'has to be invented *ex novo*', in an act of creative imagination as much as of rational analysis. Indeed, Peirce himself notes that at times 'abductive inference shades into perceptual judgment without any sharp line of demarcation between them'.[81] For this reason, we find Peirce using a variety of images and concepts to articulate what he means by abduction—such as *pattern recognition*, in which a confused tangle of things is made intelligible; the *interrogation* of a system in order to disclose its structures; and developing an *instinct* for the best explanation of phenomena.[82]

Peirce noted that the empirical evidence may suggest a number of possible abductions, forcing clarification of how the preferred abduction is to be identified. Critically, Peirce argued that abduction was fundamentally innovatory and creative, generating new ideas and insights in response to 'surprising facts'. Peirce further suggested that the human mind appears to have an instinctive capacity to relate to nature. There is an innate resonance between the human mind and nature, in that the mind has 'a natural bent in accordance with nature'.[83] Peirce sees this 'insight' as an instinctive capacity, not to

---

[77] Peirce, *Collected Papers*, vol. 4, 541 n. 1.
[78] Peirce, *Collected Papers*, vol. 2, 96.
[79] Peirce, *Collected Papers*, vol. 5, 171.
[80] Peirce, *Collected Papers*, vol. 5, 181.
[81] Peirce, *Collected Papers*, vol. 5, 181.
[82] See Hookway, 'Interrogatives and Uncontrollable Abductions;' idem, *Truth, Rationality, and Pragmatism*, 21–43.
[83] Peirce, *Collected Papers*, vol. 6, 478.

be confused with the 'powers of reason', but rather belonging to 'the same general class of operations to which Perceptive Judgments belong'.[84] This propensity may be the outcome of nurture as much as nature, representing embedded patterns of thinking as much as inborn instincts. Where others held that the emergence of novel scientific ideas was essentially random, not governed by any discernible logic,[85] Peirce holds that there is some innate human propensity to find its way to the right abduction, while leaving open how this capacity to explain might itself be explained.

Peirce's abductive approach to scientific explanation received little attention until several decades after his death, and suffered from misunderstanding on the part of his interpreters.[86] It is, however, now firmly established as a mode of thinking which is appropriate in certain contexts, including the natural sciences.

ABDUCTION IN CHRISTIAN THEOLOGY

As we noted earlier, the philosopher Charles Peirce was instrumental in developing abductive modes of thought, and demonstrating their resonance with scientific reasoning. This form of reasoning is not yet widely encountered within Christian theology, although there are indications that its considerable theological potential is coming to be appreciated.[87] For this reason, we shall focus in this section on Peirce himself, who applied this approach to theological questions, developing his own distinct approach to the rationality of religious belief.

Peirce developed an argument for the existence of God which he framed within the practice of 'musement'—an imaginatively playful reflection on the content and structures of the world, unfettered by

---

[84] Peirce, *Collected Papers*, vol. 5, 173.
[85] Popper, *The Logic of Scientific Discovery*, 20–1.
[86] McKaughan, 'From Ugly Duckling to Swan', 447–54.
[87] For example, Ben Quash's application of this mode of thinking to pneumatology, and Peter Ochs's account of its role in biblical interpretation: Quash, *Found Theology*, 208–26; Ochs, *Peirce, Pragmatism and the Logic of Scripture*.

rules or laws.⁸⁸ Peirce's thought is here guided by his theory that there is 'latent tendency toward belief in God' within every human being.⁸⁹ This being so, the thought of God is bound to arise—not as a result of any formal logic or reasoning process, but more through the process of free imaginative play.⁹⁰

In the Pure Play of Musement the idea of God's reality will be sooner or later to be found an attractive fancy, which the Muser will develop in various ways. The more he ponders it, the more it will find response in every part of his mind, for its beauty, for its supplying an ideal of life, and for its thoroughly satisfactory explanation of his whole threefold environment.

Concerns might easily be raised here, not least because, like Kant before him, Peirce seems curiously inattentive to the social and cultural embeddedness of the process of reflection. Surely musing is shaped, at least to some extent and in some manner, by culture and tradition?

Yet Peirce's point is perhaps misunderstood. Abduction is, at least in part, about an imaginative questing for the best explanation of otherwise puzzling observations. For Peirce, an abductive reasoning strategy is called for precisely because the idea of God derives from 'immediate experience', and hence cannot adequately be accommodated within alternative models of reasoning.⁹¹ There is an interesting parallel here with the biologist Simon Conway Morris's reflections on the phenomenon of convergent evolution. Why does the evolutionary process, so often thought of as shaped by random forces and radical contingency,⁹² end up converging on certain specific forms? Conway Morris argues that the number of evolutionary endpoints is limited, and argues for the predictability of evolutionary outcomes, not in terms of genetic details but rather their broad phenotypic manifestations. Convergent evolution is to be understood as 'the recurrent tendency of biological organization to arrive at the same solution to a particular need'.⁹³

---

[88] Peirce, *Collected Works*, vol. 6, 458. See further Clanton, 'The Structure of C. S. Peirce's Neglected Argument for the Reality of God'.
[89] Peirce, *Collected Works*, vol. 6, 487.
[90] Peirce, *Collected Works*, vol. 6, 465.
[91] See the discussion in Ejsing, *Theology of Anticipation*, 148–53.
[92] Gould, *The Structure of Evolutionary Theory*, 1019–20.
[93] Conway Morris, *Life's Solution*, xii.

For Conway Morris, the phenomenon of convergent evolution reveals the existence of stable regions—'islands of stability'—in biological space.[94] Evolution regularly appears to 'converge' on a relatively small number of possible outcomes: 'the evolutionary routes are many, but the destinations are limited'.[95] Even an essentially random search process will end up identifying stable outcomes in biological space.[96]

Peirce seems to develop a similar line of thought. The process of 'musement' is a creative search engine, enabling an imaginative playing with possible explanations of the rich world of observation and experience, which converges on the notion of God precisely because it is to be seen as an island of conceptual stability. As we, for example, walk beneath a night sky, contemplating the 'stars in the silence', we find ourselves playfully meditating, not according to fixed logical rules but in an imaginatively rich and unbounded manner, during which the 'idea of there being a God' constantly surfaces for consideration.[97] Peirce offers an unexplained explanation of this convergence of such musement on the idea of God: a 'latent tendency toward belief in God' within human beings.[98] Although Peirce does not make such a move, this 'tendency' can be located within the broader explanatory framework offered by a Trinitarian theological schema.[99]

As noted earlier, relatively few Christian theologians to date have explicitly adopted abductive modes of reasoning. However, such modes of thought can easily be identified in the writings of theologians, even if they are not specifically identified in this way. For example, C. S. Lewis's assertion of the rationality of religious belief in *Mere Christianity* (1952) makes use of what are clearly abductive strategies. Lewis's 'argument from desire' (141–3) is not a deductive argument for the existence of God, but is essentially an

---

[94] Conway Morris, *Life's Solution*, 127.
[95] Conway Morris, *Life's Solution*, 24.
[96] Conway Morris offers the example of the discovery of Easter Island as an example of such an inevitable discovery: Conway Morris, *Life's Solution*, 19–21.
[97] Peirce, *Collected Works*, vol. 6, 501.
[98] Peirce, *Collected Works*, vol. 6, 487.
[99] For an explicit engagement with Peirce on this point, see Robinson, *God and the World of Signs*, 308–12. More generally, see Feingold, *The Natural Desire to See God According to St. Thomas and His Interpreters*; Janz, *God, the Mind's Desire*.

abductive argument, primarily for the existence of Heaven or 'another world', and secondarily for the existence of God.

This brief account of the three main reasoning processes deployed in science and Christian theology shows how the three modes of reasoning traditionally designated as abduction, induction, and abduction are deployed in both science and Christian theology. The differences are primarily concerned with their intellectual substrates, rather than the rational processes. While both assume that it is possible to make sense of the world, within certain limits, the natural sciences tend to focus on the world of nature, while Christian theology focusses on nature and history, especially certain episodes within history which can be regarded as mediating or embodying divine disclosure.[100]

---

[100] Collins, *Trinitarian Theology, West and East*, 65–88.

7

Complexity and Mystery

The Limits of Rationality in Science and Religion

What happens if the human mind is confronted with something that is so vast that it is incapable of accommodating it? Can this incapacity to accommodate be considered an indication of irrationality? Or is it better seen as an indication of the limits of human reason, an intimation of the constraints placed on our capacity to engage and represent our world, and an accompanying plea for epistemic humility? These are not new questions, and have been discussed since the pre-Socratic period in philosophy. Yet they arguably have become of greater significance in the relatively recent past, partly because of the interest in the categories of both the 'irrational' and 'mysterious' in both religion and the natural sciences.

MYSTERY AND IRRATIONALITY

In his *Idea of the Holy* (1917), Rudolf Otto developed the concept of the numinous as a means of expressing what he considered to be the 'irrational' aspects of the holy or sacred, which he held to be foundational to religious experience and existence.[1] For Otto, the numinous can be understood as an experience of a mysterious terror and awe (*mysterium tremendum et fascinans*) in the presence of that which is 'totally other' (*das ganz Andere*), and is thus not capable of being

---

[1] Otto, *Das Heilige*. For analysis, see Schüz, *Mysterium Tremendum*, 98–297.

expressed directly using human language. Otto distinguished various aspects of the numinous, including the generation of a sense of awe (*das Schauervolle*), a sense of being overwhelmed (*das Übermächtige*), and an experience of energization (*das Energische*). Otto's conception of an irrational or numinous aspect of religion that lies beyond conceptual description and is accessible only through experience has proved significant in the study of religion; it is, however, of particular importance to any serious attempt to reflect on the religious category of mystery. While the qualities that Otto associates with the numinous can be linked with the human experience of vast or frightening natural phenomena—such as violent thunderstorms—they nevertheless have a particular association with religious contexts.

Otto is careful to avoid the suggestion that irrational means rationally deficient; his judicious use of complementary terms such as 'non-rational' or 'trans-rational' nuances his discussion, indicating that the mystery at the heart of religion is something that overwhelms and saturates human rational capacities. By holding the 'rational and non-rational elements of religion together in creative tension',[2] Otto avoids reducing religion to the rational spirituality of the Enlightenment.

Related trends can be seen in early Greek philosophy. Although many have presented classical Greek culture as a manifesto of rationalism, perhaps even as a precursor of the Enlightenment,[3] there are good reasons for suspecting the entanglement of philosophical reflection and the mystery cults,[4] pointing to a more complex notion of 'reason' than some might allow. It is clear that the Greek writers of the Archaic Age (*c.*750–480 BC) and the Classical Age (*c.*480–323 BC) were aware of the limitations of reason in grasping the complexity of the world, acknowledging that some aspects of its behaviour seemed to elude rational analysis.[5]

---

[2] Raphael, *Rudolf Otto and the Concept of Holiness*, 22. Mircea Eliade locates *Das Heilige* against a backdrop of 'irrationalistic philosophies and ideologies' to emerge after the First World War: Eliade, 'The Quest for the "Origins" of Religion', 162.

[3] For example, see Israel, *Enlightenment Contested*, 3.

[4] Riedweg, *Mysterienterminologie bei Platon, Philon, und Klemens von Alexandrien*; Martín-Velasco and Blanco, eds, *Greek Philosophy and Mystery Cults*; Kingsley, Peter. *Ancient Philosophy, Mystery, and Magic*, 71–132.

[5] Dodds, *The Greeks and the Irrational*, 254. For a wider discussion, see Elster, *Ulysses and the Sirens*; idem, *Ulysses Unbound*.

[The philosophers] who created the first European rationalism were never—until the Hellenistic Age—'mere' rationalists: that is to say, they were deeply and imaginatively aware of the power, the wonder and the peril of the Irrational. But they could describe what went on below the threshold of consciousness only in mythological or symbolic language; they had no instrument for understanding it, still less for controlling it.

The category of the 'irrational', however, is inadequate as an account of human responses to the world which lie beyond the realm of the rational. Beauty and wonder, for example, are difficult to accommodate within the highly restrictive binary proposed by the rational–irrational framework. For Aristotle, wonder was a gateway experience, itself neither rational nor irrational, for the intellectual exploration of the universe. Part of that task of investigation is governed by the need *sozein ta phainomena*[6]—to respect and safeguard the deliverances of observation and experience—while at the same time realizing that these point beyond themselves, often opening up deeper questions which call out for theoretical responses.[7]

More recently attention has focussed on the experience of awe—a feeling induced by the experience of vastness that requires some sort of mental accommodation to overwhelming new information. Recent studies in the psychology of awe have located its origins as lying in the fundamental inability of human cognitive processes to cope with the phenomenon of vastness, which compel us to expand our understanding of the world to accommodate this new information. Examples of such vast phenomena include the night sky, physical landscapes, and intellectual systems—such as Marxism or Christianity. Studies suggest that an experience of awe creates a new receptivity towards increasing understanding, thus offering a powerful stimulus to the scientific engagement with nature. The fundamental inability of the human mind to take in the vastness of a conceptually irreducible nature inevitably subverts the adequacy of or any claim to finality on the part of all attempts to categorize or represent it. It is not difficult to see how Whitehead's cautionary remarks about science might find

---

[6] Aristotle, *De Caelo*, 293a25; 296b6; 297a4. Contra Duhem, this phrase does not need to be interpreted in an instrumentalist manner: Duhem, *Sauver les apparences*, 13–37.

[7] Miller, *In the Throe of Wonder*, 3.

a wider application: 'Should we not mistrust the jaunty assurance with which each generation believes it has at last got the concepts with which to make sense of the world?'[8]

There are clearly affinities here with the notion of a 'mystery'. Although many forms of rationalism tend to portray the category of 'mystery' in terms of a crudely disguised irrationality or an evaded opportunity for rational analysis, it is better understood as something which is too great for the human mind to take in fully, often thereby forcing an appeal to the imagination, rather than the reason. The American theoretical physicist Richard Feynman (1918–88) thus argued that the human imagination often finds itself 'stretched to the utmost, not, as in fiction, to imagine things which are not really there, but just to comprehend those things which are there'.[9] An expanded vision of reason, supplemented by the imagination, is often required to capture the vision of a grand theory—whether scientific, political, or theological.

The concept of 'mystery' is thus not to be understood in terms of a crude contradiction of human rationality, but as something which calls into question the capacity of the human mind to gain a *full* grasp of our complex universe, thus exposing the limits of human rationality. This point is clearly recognized by Richard Dawkins, who remarks that the human mind, seemingly adapted to cope with the natural threats and opportunities upon which our survival arguably depends, struggles to take in the vastness of our mysterious universe. 'Modern physics teaches us that there is more to truth than meets the eye; or than meets the all too limited human mind, evolved as it was to cope with medium-sized objects moving at medium speeds through medium distances in Africa.'[10] Most evolutionary theorists would agree; the human mind developed in order to ensure our survival, rather than investigate the deep structure of the universe.

Dawkins's explicit reference to the 'all too limited human mind' might cause anxiety in some quarters; others, however, will see this as a welcome and appropriate acknowledgement of the limits of the human mind, and of the difficult questions this raises for situations or

---

[8] Whitehead, *The Concept of Nature*, 104.
[9] Feynman, *The Character of Physical Law*, 127–8.
[10] Dawkins, *A Devil's Chaplain*, 19.

phenomena which lie on its boundaries. The concept of 'mystery' does not legitimate the irrational, but highlights the possibility that we might restrict ourselves to the world of what can be proved to be 'rational'—and in doing so, needlessly confine ourselves within an intellectual prison of our own making. Max Weber spoke of this capacity of rationalism to enfold and trap human existence in systems based purely on teleological efficiency, rational calculation, and control as a *stahlhartes Gehäuse*, often rendered into English as an 'iron cage', but better translated as a 'shell that is as hard as steel'—and thus impervious to challenge and assault.[11] Yet it requires to be questioned, in terms of both its grounds and its outcomes.

A receptivity towards mystery creates the possibility of conceptual enlargement,[12] potentially liberating us from the limits of Weber's shell of steel. This theme is prominent in the writings of Giambattista Vico, who argued that the human mind can only fully grasp what human beings have themselves constructed, and is challenged and humbled when it attempts to cope with the vastness of the created order. Vico's use of the term 'mystery' is not easy to define, in that it occupies intellectual territory bordering the regions of the deliberately concealed, the intellectually impenetrable, and the hermeneutically polyvalent.[13] For Vico, the world of nature could be known fully only by God, in that God created this natural order; human beings ought therefore to focus on their own creations—such as the world of culture—which they could expect to grasp more fully.[14] Vico's approach thus suggests that both the natural sciences and Christian theology must struggle to gain even a partial grasp of the greater reality which is the object of their study, since both nature and God lie beyond human creation or control.

Many theologians would disagree with such a statement, arguing that this fails to do justice to the Christian view that God wills to be disclosed, whereas nature is open to our unaided investigations.[15] Yet Vico's approach can easily be reframed within the classical Renaissance notion of the 'Two Books'—the created order itself,

---

[11] Baehr, 'The "Iron Cage" and the "Shell as Hard as Steel"'.
[12] Kidd, 'Receptivity to Mystery'.
[13] Mazzotta, *The New Map of the World*, 115–39.
[14] Vico, *La scienza nuova*, 231–2.   [15] Torrance, *Theological Science*, 299.

and the Christian Bible, both of which demand engagement and interpretation.[16] This naturally opens up the question of how the natural sciences and Christian theology envisage and explore the concept of 'mystery'.

MYSTERY IN SCIENCE

The term 'mystery' is regularly used within the natural sciences in the general sense of 'something that is presently not understood'—but which, of course, might well be comprehended in the future, as new evidence accumulates and new theoretical models are developed. In its non-religious sense, the term essentially designates the domains of the uncomprehended and the unexplained. Most scientific writers tend to see the idea of mystery as a temporary staging post in the narrative of scientific advance. Adopting what is in effect an attitude of promissory rationalization, a mystery is glossed as a temporary inexplicability. What is mysterious today is expected to be explained tomorrow. A good example of this approach may be seen in Charles Darwin's explanation of the phenomenon of biological diversity, first set out in his *Origin of Species*.

For Charles Darwin, the question of the historical origin of species was the 'mystery of mysteries'.[17] The phrase was not new to Darwin, who rather borrowed it from the astronomer Sir John Herschel, who used it to refer to 'the replacement of extinct species by others'.[18] The solution to a mystery lies in the discovery of a higher-order theory that allows what presently seems incomprehensible or incoherent to be seen in a new way. Darwin's answer to Herschel's riddle was to find a theory which made intelligible what otherwise might seem mysterious—the theory of descent with modification by natural selection.

---

[16] Mews, 'The World as Text'.
[17] Darwin, *On the Origin of the Species by Means of Natural Selection*, 1.
[18] See Herschel's 1836 letter to Charles Lyell, published as an appendix to Charles Babbage's *Ninth Bridgewater Treatise* (1837): Cannon, 'The Impact of Uniformitarianism'.

For Darwin, a good theory allowed things to be seen in a new way, so that patterns of connections and continuities could be discerned beneath the surface of apparent happenstance and coincidence. Such a theory imposed a rational net over the complexities of experience. His theory of natural selection illuminated the evidence, leading to a growing sense that the previously inexplicable had, in fact, a plausible explanation. Certain 'mysteries' thus cease to be mysterious when they are illuminated by an informing theory, which in effect generates an intellectual framework within which they are rendered intelligible or predictable.

Yet other mysteries remain. For some, the riddle of dark energy is now the most profound mystery in all of science. Earlier scientists—such as Galileo, Kepler, and Newton—thought of as the actual universe as essentially coextensive with the observed universe. It is today thought that the observable universe represents in reality only four per cent of what really exists. Twenty-three per cent of the universe is now thought to consist of dark matter, and seventy-three per cent of dark energy.

So why should rational people believe in 'dark matter', when it is invisible? Dark matter is a hypothesis—a postulated form of matter which would explain a number of otherwise puzzling astronomical observations. Although dark matter cannot be directly observed, its existence and properties are inferred from its gravitational effects, such as the motions of visible matter and gravitational lensing. Like Newton's concept of gravity, an unobserved and unobservable hypothetical entity is invoked to explain what can be observed.

So does the notion of mystery, as used within the natural sciences, now mean little more than that which is at present scientifically inexplicable, but is expected to be explained in the future? Many might agree; yet there is a persistent note of caution sounded within at least some sections of the scientific community concerning its explanatory achievements. It is still profoundly puzzling why the deep structures of the universe can be represented mathematically. As Eugene Wigner suggested, 'the miracle of the appropriateness of the language of mathematics to the formulation of the laws of physics is a wonderful gift which we neither understand nor deserve'.[19]

---

[19] Wigner, 'The Unreasonable Effectiveness of Mathematics'.

Sometimes abstract mathematical theories that were originally developed without any practical application in mind later turn out to be powerfully predictive physical models.[20] For Wigner, this was a mystery that called out for an explanation.

Werner Heisenberg is one of many who argue that a good scientific theory would aim to 'do justice to every new experience, to every accessible domain of the world'.[21] Yet while Heisenberg celebrated the capacity of scientific theories to develop conceptualities adapted to the domain of reality under investigation within a specific discipline, he nevertheless noted that they were challenged by the 'bottomless depth' and 'impenetrable darkness' of the universe, and the shortcomings of the human intellectual struggle to find a language adequate to engage and represent this.[22]

For Heisenberg, there is a sense of mystery about our universe, in that every scientific advance simply opens up new questions, often calling into question the capacity of human minds and language to cope with the external reality that we call the universe. 'Every time when there is an understanding of a new reality, their sphere of validity appears to be pushed yet one more step into an impenetrable darkness that lies behind the ideas language is able to express.'[23] The concept of mystery can thus be seen to articulate the failures of human language to give an adequate description of the complex granularity of our universe.

Yet the growing influence of scientism on popular culture has led to resistance to the notion of mystery as a valid scientific category. As we have seen, scientism affirms the 'exclusive sufficiency' of natural scientific descriptions of the world, so that a 'mystery' is recategorized in terms of that which has not yet been explained by science. Whereas a 'mystery' in the strict sense of the term points to a vast and conceptually irreducible reality, from this scientistic stance it is seen simply as a transient feature of the human quest to make sense of the world, rather than as an irreducible and ineliminable aspect of that

---

[20] For examples and discussion, see Mario Livio, *Is God a Mathematician?* New York: Simon & Schuster, 2009.
[21] Heisenberg, *Die Ordnung der Wirklichkeit*, 38–52.
[22] Heisenberg, *Die Ordnung der Wirklichkeit*, 44.
[23] Heisenberg, *Die Ordnung der Wirklichkeit*, 44.

quest. Whereas philosophers of science such as Bas van Fraassen speak of a stance of 'abiding wonder' at the world,[24] or a sense that the world is not exhausted by scientific description, scientism depicts this as merely reflecting the limits on our present understanding of the world, which will be overcome with the progress of time and research.[25]

Albert Einstein is perhaps one of the most significant scientists to reflect on the notion of mystery. Although he was not a mystic in any meaningful sense of the term,[26] Einstein was clear that a sense of 'the mysterious (*das Geheimnisvolle*)' was the source of all true art and science.[27] Possibly picking up on Otto, Einstein suggests that the 'experience of mysteriousness (*das Erlebnis des Geheimnisvollen*)', perhaps mingled with fear, is reflected in religion. Yet Einstein does not equate 'the mysterious' with 'the irrational', seeing it rather as a gateway experience which opens the way to a rational understanding of the world, to the extent to which this is possible. 'What I see in nature is a magnificent structure that we can only grasp imperfectly.'[28] Einstein however, saw his recognition of the merits of a receptivity towards mystery as leading not to an irrational mysticism, but to a yearning to grasp nature more fully, if still incompletely and inadequately.

MYSTERY IN CHRISTIAN THEOLOGY

Earlier in this volume, we have highlighted the explanatory capaciousness of the Christian faith. This does not, however, mean that the Christian faith offers the clarity of a solution to a puzzle, in that

---

[24] van Fraassen, *The Empirical Stance*, 47–8.
[25] Cooper, 'Living with Mystery', 6.
[26] Jammer, *Einstein and Religion*, 125–7.
[27] Einstein, *Mein Weltbild*, 420.
[28] Dukas, *Albert Einstein*, 132: 'Was ich in der Natur sehe, ist eine großartige Struktur, die wir nur sehr unvollkommen zu erfassen vermögen ... Dies ist ein echt religiöses Gefühl, das nichts mit Mystizismus zu schaffen hat.' This quote dates from 1954 or 1955.

God is seen as something—someone—who resists reduction to rational commonplaces.[29]

At [the] heart [of the Christian faith] is the understanding of Christ as the divine *mysterion*: an idea central to the epistles of the Apostle Paul. This secret is a secret that has been told; but despite that it remains a secret, because what has been declared cannot be simply grasped, since it is *God's* secret, and God is beyond any human comprehension.

This notion plays an important role in the New Testament epistles, designating a distinct aspect of the Christian revelation[30]—the 'mystery that has been hidden throughout the ages and generations but has now been revealed' (Colossians 1:26).

For later writers such as Maximus the Confessor, the term 'mystery' fundamentally designated the notion of the conceptual immensity or ontological vastness of God. A mystery is resistant to interpretative closure or intellectual reduction, ultimately transcending any attempt at a limiting definition—precisely because this limits what must be allowed to remain open. The English theologian Charles Gore highlighted the importance of this point, noting the limits of language to comprehend the mystery of the divine:[31]

Human language never can express adequately divine realities. A constant tendency to apologize for human speech, a great element of agnosticism, an awful sense of unfathomed depths beyond the little that is made known, is always present to the mind of theologians who know what they are about, in conceiving or expressing God.

Although the concept of 'mystery' was used extensively by patristic writers,[32] and remains important for many theologians today,[33] a sense of unease remains over the possible elision of the category of 'mystery' with that of the 'irrational'. Where some elements of the Enlightenment insisted that the investigation of reality should be

---

[29] Louth, *Origins of the Christian Mystical Tradition*, 205. See also Louth, *Discerning the Mystery*.
[30] Lang, *Mystery and the Making of a Christian Historical Consciousness*, 9–128.
[31] Gore, *The Incarnation of the Son of God*, 105–6.
[32] See the richly documented discussion in Fiedrowicz, *Handbuch der Patristik*, 642–59.
[33] Louth, *Discerning the Mystery*, 132–47.

capable of being expressed using clear and distinct ideas, the concept of mystery is profoundly and intrinsically resistant to any such conceptual closure. It is not difficult to see how the resistance of such a mystery to such forms of rational interrogation came to be seen by some as an indication of its fundamental irrationality, rather than its necessary resistance to such reductionist analysis.

The French theologian and philosopher Gabriel Marcel (1889–1973) helpfully drew a distinction between 'problems' and 'mysteries'.[34] For Marcel, the world of problems is the domain of science, rational enquiry, and technical control. We live in a 'broken world (*un monde cassé*)', which is resistant to a disinterested total comprehension. This 'broken world' is 'riddled with problems' on the one hand and, on the other, is 'determined to allow no room for mystery'.[35] For Marcel, a problem is something which can be viewed objectively, and for which we can find a possible solution. A mystery, however, is something which we cannot view objectively, precisely because we cannot separate ourselves from it.[36]

A problem is something which I meet, which I find completely before me, but which I can therefore lay siege to and reduce. But a mystery is something in which I am myself involved, and it can therefore only be thought of as a sphere where the distinction between what is in me and what is before me loses its meaning and initial validity.

While problems can give rise to universal or generalized solutions, mysteries simply do not admit such generalized solutions. To ask about the meaning of life as if it were a mere object to be analysed disinterestedly is in effect to act as if the answers have no bearing on our own existence.

Life, according to Marcel, is thus not a problem to be solved but a mystery to be lived. The existence of suffering, for example, is thus to be seen as a mystery that can never be fully grasped, rather than as an intellectual problem that can be mastered and subdued.[37] We find ourselves unable to place ourselves wholly outside a mystery in order

---

[34] See especially Marcel, *Being and Having*.
[35] Marcel, *The Philosophy of Existentialism*, 12. For a good discussion, see Tobin, 'Toward an Epistemology of Mysticism'.
[36] Marcel, *Being and Having*, 117.
[37] See, for example, the discussion in Labbé, 'La souffrance: problème ou mystère'.

to investigate it, in that there is no objective stance, no Archimedean standpoint, from which I can observe it. In the case of suffering or evil, we have to come to terms that these are things we *experience*, not simply observe.[38]

Evil which is only stated or observed is no longer evil which is suffered: in fact, it ceases to be evil. In reality, I can only grasp it as evil in the measure in which it *touches* me—that is to say, in the measure in which I am *involved*... Being 'involved' is the fundamental fact.

Marcel's ideas were developed in a more explicitly theological manner by the English philosophical theologian Austin Farrer, who defines the realm of the problematic as 'the field in which there are right answers'. The realm of mystery, however, involves engagement with reality at such a level that it cannot be investigated in terms of 'determinate and soluble problems'. The theologian, Farrer argues, is not faced with the limited and manageable relation which arises between a conceptual instrument and physical objects; rather, we are confronted 'with the object itself, in all its fullness', and this object presents itself, not as 'a cluster of problems but as a single though manifold mystery'.[39] It is tempting to reduce a mystery to a set of problems, on the basis of the mistaken belief that the mystery is the mere sum of the individual (soluble) problems.

Marcel's approach involves the distinction between a problem and a mystery, not the rejection of the category of the problematic as such. His position is rather that problem-oriented approaches have their own distinct domains of competence—for example, in relation to scientific understanding and technological advance. Yet the difficulty is that some allow a problem-solving approach to the world to become imperialistic, holding that such an approach alone has the right to judge all knowledge and truth on the basis of 'criteria appropriate only to the aspect of the objective'.[40] For Marcel, subjective involvement and participation are a precondition for gaining an authentic and distinctive knowledge of a mystery.

---

[38] Marcel, *Philosophy of Existentialism*, 19.
[39] Farrer, *The Glass of Vision*, 72. It is interesting to note that Farrer's wife, Katherine, translated *Être et avoir* into English.
[40] Keen, *Gabriel Marcel*, 19.

So how does Marcel understand rational reflection to fit into his framework? Marcel is clear that such reflection is necessary and appropriate in coming to know both problems and mysteries; nevertheless, he argues that a distinction must be drawn between primary reflection, which is the type of rational thought involved in problem-solving, and secondary reflection, by which we know a mystery.[41] Primary reflection is thus analytic, in that it seeks to achieve clarity about the world of abstraction, objectification, and verification; whereas secondary reflection is synthetic, in that it seeks to frame a wider and richer understanding of the meaning of human existence by integrating its many facets. Whereas primary reflection seeks to dissolve the unity of experience in an act of analysis, secondary reflection is 'recuperative', in that it seeks to recapture or reclaim the unity of experience.

The analysis presented by Marcel and Farrer points to the perennial nature of the theological task, in that each generation is called upon to wrestle with a mystery, knowing that it possesses a certain inexhaustibility which cannot be grasped or fully comprehended by any one writer or era. The problematical is the domain of science and rational enquiry. Once a problem is solved there is no more interest in it. A mystery, however, challenges, refreshes, and reinvigorates the theological task, not least through the expectation that fresh light has yet to break forth from mysteries which have been wrestled with by previous generations. The process of wrestling with a mystery thus remains open, not closed. What one generation inherits from another is not so much definitive answers as a shared commitment to the process of wrestling.

Similarly, our minds struggle to even begin to cope with the immensity and majesty of God. Christian theology has long recognized that it is impossible for us to represent or describe God adequately using human language. The sheer vastness of God causes human images and words to falter, if not break down completely, as they try to depict God fully and faithfully. A mystery is not something that is contradicted by reason, but is rather something that exceeds reason's capacity to discern and describe—thus transcending, rather than contradicting, reason. As the Puritan writer Richard Baxter once

---

[41] Sweetman, *The Vision of Gabriel Marcel*, 55–60.

remarked, we may well *know* God; yet to *comprehend* God fully lies beyond our capacity.[42] For Marcel, a mystery can be 'known'; it cannot, however, be fully and objectively comprehended. 'The mysterious is not the unknowable, the unknowable is only the limiting case of the problematic.'[43]

To speak of some aspect of nature or God as a 'mystery' is not to attempt to shut down the reflective process, but to stimulate it, by opening the mind to intellectual vistas that are simply too deep and broad to be fully apprehended by our limited human vision. We can only cope with such a mystery either by filtering out what little we can grasp, and hope that the rest is unimportant; or by reducing it to what our minds can accommodate and thus to the rationally manageable. Inevitably, both these strategies end up distorting, disfiguring, and misleading, presenting God as something that one can know about, rather than one who is known, and by being known, is found to be transformative and regenerative.

Augustine of Hippo stressed the limits of our ability to capture God in neat formulae. 'If you think you have grasped God, it is not God you have grasped'—*si comprehendis non est Deus*.[44] Anything that we can grasp fully and completely *cannot* be God, precisely because it would be so limited and impoverished if it could be fully grasped by the human mind. If you can get your mind around it, it is not God, but is rather something else that you might incorrectly *think* is God. It is easy to create a god in our own likeness—a self-serving human invention that may bear some passing similarity to God, but falls far short of the glory and majesty of the God who created and redeemed the world.

THE TRINITY AS MYSTERY

Some suggest that mystery is merely a superstitious person's way of referring to an irrationality. For many outside the Christian community, the doctrine of the Trinity is a classic instance of the irrationality

---

[42] Baxter, *Practical Works*, vol. 13, 29.
[43] Marcel, *Being and Having*, 118.
[44] Augustine, *Sermo* 117.3.5. Cf. van Bavel, 'God in between Affirmation and Negation According to Augustine'.

of faith; for Christian theologians, however, it is an equally classic example of a mystery, a vision of God which is too vast to be captured by the human mind, forcing the human mind to adapt to its contours, rather than permitting the mind to reduce this vision of God to intellectually manageable proportions. The theological notion of 'glory' articulates a fundamental theme of the Christian faith: that in the end, the human mind is not capable of accommodating the conceptual vastness of God, who overwhelms our mental capacities.

During the 'Age of Reason', the rationality of this doctrine came under radical scrutiny. Early criticisms of the doctrine tended to reflect a desire for the theological simplicity of the New Testament, and a suspicion of what was seen as a promiscuous inflation of biblical ideas, especially in scholastic theology.[45] Although Isaac Newton's anti-Trinitarianism may have reflected the increasingly rationalist outlook of his age, it seems to have been mainly a consequence of his biblical hermeneutics,[46] particularly a concern that the doctrine was ultimately idolatrous. Newton's views were, in certain ways, typical of his age. Most leading theologians of the seventeenth century seem to have held on to the doctrine of the Trinity out of respect for tradition,[47] while privately conceding that it seemed irrational in respect of both its foundations and modes of expression, and offered little in the way of spiritual or theological benefits.[48] A routine defence of this doctrine seems to have been seen as being little more than a formal expectation on the part of orthodox theologians. There was no sense that the doctrine preserved or articulated something of fundamental importance, or served as the foundation or summation of key themes of faith, or even an anticipation of the twentieth-century theologian Emil Brunner's notion of the Trinity as a 'security doctrine (*Schutzlehre*)', protecting Christian theology against deficient notions of God.[49]

---

[45] For early German anti-Trinitarianism, see Dingel and Daugirdas, *Antitrinitarische Streitigkeiten*, 3–17. For Michael Servetus's anti-Trinitarianism, see especially Sánchez-Blanco, *Michael Servets Kritik an der Trinitätslehre*, 58–60.
[46] Mandelbrote, 'Eighteenth-Century Reactions to Newton's Anti-Trinitarianism'.
[47] Rogers, 'Stillingfleet, Locke and the Trinity'.
[48] For a detailed account, see Lim, *Mystery Unveiled*.
[49] Brunner, *Dogmatik I*, 206.

The theological generations to follow Newton tended to adopt an essentially deist notion of God in their public defence of Christianity, seeing this as minimally counterintuitive to the norms of an emerging rational and scientific professional culture, and yet sufficiently orthodox to resonate with those of a still predominantly religious—in this case, predominantly Anglican—society.[50] Deism—a distinctly English phenomenon—is best seen as a rationally maximized notion of God which set out to secure cultural compliance and conformity at a time of cultural change and uncertainty.

Many Western theologians of the eighteenth—and even the twentieth—century seem to have taken the view that the 'Age of Reason' had achieved a permanent cultural hegemony, and that as a result theology had little option other than to submit to its rational norms. As late as 1977, Leslie Houlden argued that 'we must accept our lot, bequeathed to us by the Enlightenment, and make the most of it'. Few would now agree. Houlden's failure to grasp the significance of the postmodern deconstruction of Enlightenment rational norms is puzzling; his indifference to the revival of Trinitarian theology is culpable. Today, the doctrine of the Trinity is central to Christian theological discourse. Karl Barth and Karl Rahner have catalysed a major programme of theological retrieval, in which the rationality and utility of the doctrine of the Trinity have been reaffirmed.[51] The doctrine of the Trinity is seen to articulate the distinct theological logic of the Christian faith,[52] inviting a comparison and correlation with other notions of rationality, rather than leading to the assimilation of such a theological rationality to the norms of the 'Age of Reason'.

The theological notion of mystery is illuminated by the psychology of awe. The psychologists Dacher Keltner and Jonathan Haidt developed a prototype approach to the experience of awe, which has at its heart two distinctive themes—the *vastness* of nature and the cognitive

---

[50] Mandelbrote, 'The Uses of Natural Theology in Seventeenth-Century England'.
[51] See Davis, Kendall, and O'Collins, eds, *The Trinity*.
[52] For careful analysis of an early modern statement of this approach, see Burton, *The Hallowing of Logic*, 72–94. More generally, see Coakley, 'Living into the Mystery of the Holy Trinity'.

process of *accommodation*.⁵³ Vastness is here to be understood as 'anything that is experienced as being much larger than the self, or the self's ordinary level of experience or frame of reference'. It may refer simply to physical size, or to more subtle markers of vastness, such as social signs or symbolic markers. Accommodation refers to the process identified by Jean Piaget (1896–1980), professor of genetic and experimental psychology at the University of Geneva from 1940 to 1971. Piaget defined this as the process by which human mental structures undergo an adjustment in the face of the challenge posed by new experiences. Thus it would be possible to experience a sense of awe through realization of the 'breadth and scope of a grand theory', such as evolutionary theory—or the Christian vision of reality.

We propose that prototypical awe involves a challenge to or negation of mental structures when they fail to make sense of an experience of something vast. Such experiences can be disorientating and even frightening . . . They also often involve feelings of enlightenment and even rebirth, when mental structures expand to accommodate truths never before known. We stress that awe involves a *need* for accommodation, which may or not be satisfied. The success of one's attempt at accommodation may partially explain why awe can be both terrifying (when one fails to understand) and enlightening (when one succeeds).

For Piaget, human beings interact with their environment through a process of 'reflecting abstraction (*l'abstraction réfléchissante*)'. Human beings are not born with such structures, nor do they absorb them passively from their environment: they construct them through a process of interaction (which Piaget terms 'equilibration'), in which an equilibrium is achieved between assimilation and accommodation. Assimilation may be defined as the 'act of incorporating objects or aspects of objects into learned activities', whereas accommodation is 'the modification of an activity or ability in the face of environmental demands'.⁵⁴ These interact in an adaptive process which permits new information or observation to be fitted into already existing cognitive structures, leading to equilibration, in which a balance is maintained

---

[53] Dacher Keltner and Jonathan Haidt, 'Approaching Awe, a Moral, Spiritual and Aesthetic Emotion', *Cognition and Emotion* 17 (2003): 297–314.
[54] As defined by Lefrançois, *Theories of Human Learning*, 329–30.

'between assimilation (using old learning) and accommodation (changing behaviour; learning new things)'. Assimilation can thus be thought of as applying an existing scheme of thought to a new situation, initiating a process of comparison and evaluation.[55] At times, that may lead to the realization that the old way of thinking is not capable of coping with what is being observed, forcing a modification of this scheme—in other words, accommodation.

A mystery is thus something that is conceptually vast which triggers a process of accommodation. Keltner and Haidt identify the possible outcomes of a failure to comprehend an experience of something that is vast, which includes fear and disorientation. This is not, it must be emphasized, an affirmation of the irrationality of nature or of human responses to nature; it is simply a recognition of the challenges that human cognitive processes experience when confronted with something immense.

On the basis of this discussion, it seems that the concept of mystery retains validity as a framing device for approaching the human experience of vastness—whether in the form of natural landscapes, conceptual systems, or the apprehension of the universe—which overwhelms the human mind, and thus encourages reductionist or assimilationist coping strategies which in effect reduce reality to what can be managed. The concept of mystery represents a protest against such strategies, demanding the preservation of the phenomena, irrespective of how difficult they may be to grasp or comprehend.

ON BEING RECEPTIVE TO MYSTERY

It is a matter for some regret that there has been relatively little philosophical discussion of the category of 'mystery'. Yet it is not difficult to understand a reluctance to engage with this notion, given both its intrinsic resistance to definition on the one hand, and the associations, however unjustified these might be, with irrationalism or specifically religious ideas, which might be deemed to lie beyond

---

[55] Lefrançois, *Theories of Human Learning*, 335.

the legitimate scope of philosophy on the other. Yet perhaps one of the most important concerns is the tension between the category of a 'mystery' and the characteristic Enlightenment demand for clear and distinct ideas about ourselves and our world.[56] This reflects Descartes's influential principle of clear and distinct perception, as set out in his 'Third Meditation',[57] which Descartes himself regarded as supportive of belief in a good and perfect God, yet which later thinkers came to see as inimical to a Trinitarian notion of God.[58] There is a sense in which a 'mystery' is something which cannot be reduced to such well-defined ideas; to neutralize its threat to the authority of reason, it was therefore necessary for rationalists to classify this as an irrational belief, rather than something which exposed the limits of reason.

David E. Cooper and John Cottingham should be noted here as examples of significant contemporary philosophical voices which affirm the central importance of 'experiences of mystery' or 'intimations of the transcendent' for religion, even though they diverge somewhat in their understandings of quite what form such experiences and intimations might take.[59] Cottingham explores the notion of religion—or being religious—in terms of living in responsive awareness of the 'mystery of existence'. For Cooper, we have to face up to reality's being 'ineffable and mysterious', in that 'no account of the world' could count as a 'description of reality as such', since any such descriptions would ultimately reflect or be shaped by the prejudices, purposes, practices, and perspectives of those who devise them.

A point which emerges from the recent work of both Cooper and Cottingham is the need to come to terms with the limited hold which the human mind has on reality.[60] Although they express this in different ways, there are some common themes which emerge from their reflections. 'Mystery' refers to or involves the recognition of something that is 'beyond the human', whether this is understood *metaphysically* as a transcendent reality that resists reduction to

---

[56] A topic explored in detail in Lifschitz, *Language and Enlightenment*. See also Kenshur, *Dilemmas of Enlightenment*, 97–100.
[57] Marion, 'The General Rule of Truth in the Third Meditation'.
[58] Powell, *The Trinity in German Thought*, 60–3.
[59] Cooper, *The Measure of Things*; idem, 'Living with Mystery'; Cottingham, 'Religion and the Mystery of Existence'.
[60] For a good discussion, see Kidd, 'Receptivity to Mystery'.

Complexity and Mystery

human terms, or *epistemologically* as a realization, however reluctant, that the world simply cannot be reduced to the humanly comprehensible without intellectual distortion or degradation.

MYSTERY: AN INVITATION TO DEEPER REFLECTION

Some critics of religion have suggested that an appeal to mystery represents an illegitimate attempt to shut down reflection on the nature of the universe, or a perverse celebration of human irrationality. Others, however, see this as a necessary and proper recognition of the complexity of our universe, and the limits placed upon human cognition. Colin McGinn and others have developed an approach which might be termed 'epistemic mysterianism'.[61] On this view, what makes a mystery intractable for us is primarily the poverty of our epistemic capacities, not the way the world is. While there are problems with such an approach, it nevertheless highlights the potential implications of the limits placed on human cognitive processing.

Yet when properly understood, the theological category of 'mystery' simultaneously draws us to itself by its depth and luminosity, while frustrating our capacity to dissect it, and reduce it to manageable components. This point was made with particular clarity by Leonardo Boff:[62]

Seeing mystery in this perspective enables us to understand how it provokes reverence, the only possible attitude to what is supreme and final in our lives. Instead of strangling reason, it invites expansion of the mind and heart.

A mystery invites engagement, yet resists closure. Part of the rationalist discontent with a mystery lies in its quest for the closure of such questions in clear and distinct Cartesian terms. Yet such a closure is ultimately 'the imposition of fixity on openness',[63] limiting our capacity to revisit a mystery and reflect further on its significance and meaning. The human desire for a firm answer to a question and an

---

[61] Kriegel, 'The New Mysterianism and the Thesis of Cognitive Closure'.
[62] Boff, *Trinity and Society*, 159.   [63] Lawson, *Closure*, 5.

aversion to ambiguity is reflected in a 'seize and freeze' mentality,[64] which reflects a perceived need to foreclose a complex discussion, perhaps arising from an ideological prejudice which is resistant to evidential interrogation and challenge.[65]

To speak of aspects of our world as a 'mystery' may indeed represent a disinclination or inability on the part of some to engage seriously and profoundly with the complexity of our world, and may thus be an indicator of an implicit irrationalism. Yet there are other possibilities, perhaps the most important of which is a profound respect for the singularity of our universe, resulting in a principled refusal to rush into premature closure of something that, by its very nature, demands extended reflection, and may ultimately prove to lie beyond an innate human capacity to comprehend those possibilities fully. That, however, does not require the mystery itself nor its human interpreters to be deemed to be 'irrational'. It is simply a recognition that some aspects of our complex world may lie beyond our capacity to grasp them fully, linked to a suggestion that we recognize our rational limits, rather than force reality into a Procrustean mould predetermined by the limits of our reason.

---

[64] Kruglanski and Webster, 'Motivated Closing of the Mind'.
[65] Roets and van Hiel, 'Allport's Prejudiced Personality Today'.

8

Rational Consilience

Some Closing Reflections on Science and Christian Theology

At the end of the twentieth century, the sociobiologist E. O. Wilson reintroduced the term 'consilience' to refer to a vision of the bringing together of outcomes and insights from different fields of knowledge, while linking this notion to the Enlightenment's goal of creating a grand unified knowledge built upon a set of universal laws,[1] and to the privileged place of the natural sciences within that goal. That vision, as we have noted, ultimately rested on an unsustainable idea of a universal rationality. So can the goal of bringing together such insights from multiple facets of human creativity and reflection—including science and theology—survive the lingering death of the notion of a universal rationality? In this final chapter, I shall suggest that it can, and will offer a brief exploration of how it might be implemented—not to solve the issues, but rather to catalyse further discussion of this important intellectual and cultural theme.

Yet rational consilience can be achieved without relying on outdated Enlightenment assumptions or succumbing to the temptations of scientism. Wilson's vision of consilience rests on a reaffirmation of the Enlightenment idea of knowledge (which has become unsustainable), and the privileging of science (which is both unwarranted and unwise). Wilson himself was an advocate of intellectual colonization, arguing that the humanities and social sciences should be seen as

---

[1] Wilson, *Consilience*, 8–9.

little more than 'specialized branches of biology'.[2] The diversity of approaches evident within the natural sciences makes it clear that we cannot talk about 'science' in general, but have to recognize the distinctiveness of each scientific discipline, and its associated rational beliefs and practices. Bruno Latour's suggestion that we need to move from a 'culture of science' to a 'culture of research' is helpful in refocussing discussion on research objectives, while emphasizing the importance of interconnection between disciplines themselves, as well as the society they serve.[3] Transdisciplinarity offers a much better framework for the achievement of real consilience than the approach commended by Wilson himself.

TOWARDS A 'BIG PICTURE': A METAPHYSICAL TURN

The physicist and Nobel Laureate Eugene Wigner once remarked that science is constantly searching for the 'ultimate truth', which he defined as 'a picture which is a consistent fusion into a single unit of the little pictures, formed on the various aspects of nature'.[4] It is a powerful image, hinting at a fragmentary picture of an unassimilated accumulation of experience and observation, and raising the question of how, whether, and to what extent, a unified view of nature might be achieved. As I opened this work by reflecting on the challenges facing anyone trying to integrate—or even intertwine—our 'thoughts about the world' with our thoughts about 'value and purpose', it seems appropriate to end this work by attempting to map out a framework within which this might be attempted, if not completely accomplished. Can we achieve a unified picture? For a single grand narrative?[5] Or must we settle for a dappled world, in which each domain

---

[2] Wilson, *Sociobiology*, 547.
[3] Latour, 'From the World of Science to the World of Research?' For its further development, see Maasen, Lengwiler, and Guggenheim, 'Practices of Transdisciplinary Research'.
[4] Wigner, 'The Unreasonable Effectiveness of Mathematics'.
[5] For the theological potential of this notion, see Sandler, 'Christentum als große Erzählung'.

represents an intellectual island characterized by its own distinct and local rationality?

This latter view, recently championed by Nancy Cartwright, picks up on Otto Neurath's criticism of 'one great scientific theory' into which all the intelligible phenomena of nature can be fitted in 'a unique, complete and deductively closed set of precise statements'.[6] For Cartwright, we live and think within a 'dappled world' in which each aspect of reality has its own separate truth and its distinct 'model'. Yet while Cartwright's analysis has some important criticisms to make of those who present natural scientists as 'nomological machines' aiming to construct a complete and deductively closed set of statements, she fails to engage what, for most natural scientists, is the most obvious and plausible way of correlating different aspects of the scientific enterprise—namely, scientific knowledge as a multiply-connected web.[7] Cartwright's failure to recognize this interconnectedness is evident in her presentation of classical Newtonian mechanics, quantum mechanics, quantum field theory, and Maxwell's electromagnetic theory as logically independent and separate, when they are clearly to be seen as different yet interconnected aspects of the same physical theory. Cartwright tends to think of these theories as existing in 'cocoons'. This may well be true in the case of Freudianism; whereas epistemic approaches to scientific explanation emphasize the capacity of good theories to interconnect, and thus to break down interdisciplinary barriers. 'In the modern state of science, no discovery lives in a cocoon; rather it is built within and upon the entire interconnected structure of what we already know.'[8]

So what kind of 'interconnected structure' might this be? The central argument of this work is that any such discussion about integration of disciplinary insights or perspectives must be set in the context of an actual and legitimate plurality of methodologies and rationalities across intellectual disciplines, along with a principled refusal to accord privilege to any beyond its own domain of competency. This book has sought to subvert some older and influential discussions of the

---

[6] Cartwright, *Dappled World*, 5–7.
[7] See the important criticisms in Anderson, 'Science: A "Dappled World" or a "Seamless Web"?'
[8] Anderson, 'Science: A "Dappled World" or a "Seamless Web"?', 490–1.

rationality of faith, primarily those associated with certain sections of the Enlightenment, which portray human rationality as a universal norm, independent of culture, history, or disciplinary specificity. The best antidote to such a superficial account of rationality is to present a more nuanced and complex reading of the situation, which is attentive to empirical findings in the social and psychological sciences. Although this notion of a historically and culturally invariant rationality lingers in some quarters, it is clear that we need to consider what the implications of 'Multiple Situated Rationalities' might be for discussion of the relation between any intellectual disciplines, including the relation of the natural sciences with Christian theology.

On the basis of her social constructivist (mis)reading of some recent scientific discussions, Cartwright argues for a 'metaphysical nomological pluralism';[9] I see no persuasive reason to do this, and argue that it is better to think in terms of our world as an ontological unity which demands and deserves a methodological pluralism, which gives rise to an interconnected web of ideas and images of scientific theories. On the basis of the evidence presented in this book, I suggest that we should rather think in terms of the principled colligation of 'little pictures', creating a larger picture, a seamless web of cross-relationships—perhaps to be visualized as a panorama, which connects and coordinates these snapshots.

The quest to construct such a 'big picture' of reality is often framed in metaphysical terms. While the nature and place of metaphysics in scientific and theological reasoning remain an open question, there is sympathy for the views that the task of metaphysics is to 'explain the world' in terms of its fundamental structures, or to explore the 'ontological commitments' of well-established scientific theories.[10] The same task is also undertaken by Christian theology, which is more amenable to exploring its metaphysical aspects.[11]

Any satisfactory and healthy Christian theology simply cannot dispense with, or be constructed in isolation from, some overall metaphysical scheme or vision which somehow articulates into a rational unity [our] experience and knowledge of the world, taken in the widest possible sense.

---

[9] Cartwright, *Dappled World*, 31.   [10] Sider, *Writing the Book of the World*.
[11] Richmond, *Theology and Metaphysics*, xi.

Yet this does not mean that metaphysics determines either scientific or theological analysis; it is rather the outcome of that process of analysis, which leads to a quest for a unifying, or at least coordinating, framework that holds together coherently the multiple insights that have been gained.

The natural sciences are characterized by their empirical stance, and a principled refusal to accept predetermined theories or accounts of our world, laid down in advance by other scientists, philosophers, or theologians. Many would argue that the natural sciences make no a priori claims or assumptions, but generate their theories through a constant probing of reality through observation, experimentation, and theoretical reflection. In this sense, scientific knowledge is essentially and characteristically a posteriori—the outcome, not the precondition or presupposition, of an empirical method. No metaphysical assumptions are required for the scientific enterprise,[12] other than the arguably functional or pragmatic assumption that nature is uniform.[13]

The Vienna Circle adopted a radical empiricism which in effect limited meaningful statements to what could be verified from observation. Its failings were obvious. Karl Popper pointed out that such an empiricism 'did not exclude obvious metaphysical statements; but it did exclude the most important and interesting of all scientific statements, that is to say, the scientific theories, the universal laws of nature'.[14] Those who develop empirical stances today—such as Bas van Fraassen—are more attentive to the 'phenomenology of scientific activity', which includes a pragmatic motivation for scientists to construct theories that describe unobservable structures beyond the phenomena, given that this methodology has in the past generated empirically adequate theories.[15] An empiricist can adopt an essentially pragmatic commitment to metaphysical theories, understanding these as ways the world might be, in much the same way as a constructive empiricist has a pragmatic commitment to theoretical

---

[12] Ladyman and Ross, *Every Thing Must Go*.
[13] An assumption which is, of course, problematic in itself: see Rowan, 'Stove on the Rationality of Induction and the Uniformity Thesis'.
[14] Popper, *Conjectures and Refutations*, 281.
[15] van Fraassen, *The Scientific Image*, 80–3.

superstructures, and recognizes the value of realist interpretations of science as accounts of ways the world might be.

These lines of thought point to the possibility of an a posteriori metaphysics—a provisional way of envisaging the world, which results from the rigorous application of an empirical method on the one hand, and a willingness to hypothesize a deeper unobservable structure beyond those phenomena, within which specific scientific disciplines operate, on the other. On this view, an implicit metaphysics underlies any empirical science, in that it provides the framework within which such sciences are conceived, undertaken, and related to one another. Yet such a framework is understood as provisional and corrigible, open to revision as scientific investigation of the world continues.

A similar pattern of thought can be discerned within Christian theology. Although some theologians draw on a 'first philosophy' to lay down an a priori framework within which theology is to be undertaken, most resist this, holding that the Christian theology is not grounded in some presupposed metaphysics, but represents a way of thinking about and imagining our world which is disclosed through Scripture, and passed down and consolidated through the Christian community, which here functions as an 'epistemic community'. For most, Christian theology is ultimately grounded, not in philosophical 'first principles', but in a narrative—especially as this focusses on the history of Jesus Christ.[16] Some early Christian theologians, embedded in a Hellenistic cultural context, used the language of Greek metaphysics in developing their theology (for example, in the statements of the Council of Chalcedon); this, however, is not to be seen as an assimilation of Christology to secular Greek thought, but a strategic deployment of its vocabulary to map the contours of the Christian vision of reality, which is rooted in (and is to be evaluated on the basis of) the New Testament.[17]

Christian theology is seen as an extended reflection on essentially empirical data—primarily the narrative of Jesus Christ—with the object of developing the larger account of reality that this intimates

---

[16] See the seminal account of this process of reflection on a foundational narrative in Niebuhr, *The Meaning of Revelation*, 23–46.

[17] O'Leary, *Questioning Back*, vii–xvii.

and enfolds. This process of reflection gradually leads to the emergence of a greater vision of the world which could be described as 'metaphysical'. Christian theology aspires to articulate such a 'scheme or vision', especially in highlighting its capacity to hold together our experience of the world as a coherent whole. 'This is our first demand of religion—that it should illumine life and make it a whole.'[18]

This process of grasping a complex interconnected structure lies beyond the capacity of any one discipline, and involves an individual thinker developing a first-person mental map of reality which enables us to imagine such a coordinated reality, and begin to grasp the nature and implications of its interconnections.[19]

Reality is one and truth indivisible. Each special science aims at truth, seeking to portray accurately some part of reality. But the various portrayals of different parts of reality must, if they are all to be true, fit together to make a portrait which can be true of reality as a whole. No special science can arrogate to itself the task of rendering mutually consistent the various partial portraits: that task can alone belong to an overarching science of being, that is to ontology.

In attempting to construct such a 'big picture' of reality, we need to attempt to colligate disciplinary insights, while recognizing that no discipline in itself can hope to provide a universally acceptable rendering of a metaphysical vision like this which renders such insights into a rational unity.

In his remarkable essay 'Resuming the Enlightenment Quest', Edward O. Wilson notes the growing trend to regard the natural sciences, social sciences, and humanities as separated 'by an epistemological discontinuity, in particular by possession of different categories of truth, autonomous ways of knowing, and languages largely untranslatable into those of the natural sciences'. Yet this, he argues, can be challenged and countered by an appeal to 'phenomena bound up with the material origins and functioning of the human brain', thus providing a basis for the unity of knowledge which privileges the biological sciences.[20] Yet Wilson's approach to consilience fails to address the issue, not least because it implies that we can

---

[18] Wood, *In Pursuit of Truth*, 102.   [19] Lowe, *The Four-Category Ontology*, 4.
[20] Wilson, 'Resuming the Enlightenment Quest', 17.

infer the *content* of human thought from knowledge of the underlying brain mechanisms.[21] Where Wilson proposes a hierarchical materialist unification, based on the intellectual hegemony of the natural sciences, I instead propose to think in terms of webs of disciplinary interconnection and cross-fertilization, which avoids such privileging of specific disciplines, and is attentive and respectful to their intellectual approaches and practices.

THE COLLIGATION OF INSIGHTS

Werner Heisenberg suggested that the Copenhagen approach to quantum theory implied that we do not know nature directly, but rather indirectly, as a result of our methods of investigation.[22] Yet if we think of nature as a totality, Heisenberg's line of thought leads us to the conclusion that a multiplicity of research methods leads to a corresponding plurality of perspectives or insights, which thus require to be integrated, coordinated, or colligated in order to allow the best possible overall representation of nature.

Although Heisenberg himself does not use the term 'pluralism' in describing his own position, the position that he develops in his later writings recognizes the complexity of both the natural world and human experience, and offers an account of this which recognizes a plurality of approaches and intellectual outcomes.[23] Perhaps most importantly, Heisenberg was able to accommodate both art and religion within his overall approach, distinguishing these methodologically from the sciences, while affirming their intellectual and cultural legitimacy and distinctiveness.[24]

---

[21] Davis, 'The Importance of Human Individuality for Sociobiology'.

[22] Heisenberg, 'Die Kopenhagener Deutung der Quantentheorie', 85: 'Und wir müssen uns daran erinnern, daß das, was wir beobachten, nicht die Natur selbst ist, sondern Natur, die unserer Art der Fragestellung ausgesetzt ist.' The Copenhagen interpretation, of course, is only one of a range of options within the field: see Schlosshauer, Kofler, and Zeilinger, 'A Snapshot of Foundational Attitudes toward Quantum Mechanics'.

[23] Schiemann, 'Welt im Wandel', 310–11.

[24] See especially Heisenberg, 'Naturwissenschaftliche und religiöse Wahrheit', 348–9.

I have already referred several times in this work to Steven Rose's fable about five biologists on a picnic who offer quite distinct accounts of why a frog jumps (59; 66–7). Their five different answers reflect the different perspectives and methodologies of the five disciplines here represented; nevertheless, they are capable of being correlated and colligated, to yield a larger account of why the frog jumped than that provided by any single perspective. Yet this account is not merely additive; it *expands* our overall account of this phenomenon, presenting a whole that is greater than its constituent parts.

The 'colligation' was used by the empirical philosopher William Whewell, who conceived it as an 'act of thought' that brings together a number of empirical facts by 'superinducing' upon them a way of thinking which unites the facts, in much the same way as a string holds together the pearls of a necklace.[25] Where Whewell tended to think of colligation as the connection of *observations*, however, I shall use the term to refer to the epistemic process of constructing a 'big picture' that is capable of accommodating and interconnecting multiple *notions or insights*, drawn from across intellectual disciplines, distinguished by their operative rationalities.

This colligation neither *resolves*, nor *depends on resolving*, two significant issues. First, like any scientific theory, each of these disciplinary accounts has to be regarded as provisional and corrigible, subject to correction through the ongoing processes of theoretical refinement and evidentiary accumulation. And secondly, this colligation of perspectives does not involve the integration or reconciliation of the distinct methodologies or operational rationalities of these disciplines. Each discipline retains its own distinctive identity, while nevertheless being able to engage in a meaningful and constructive conversation with other fields of enquiry, contributing to an interconnected web of insight. This point is clearly of relevance to any attempt to colligate insights from multiple disciplines, characterized by quite different operative rationalities—such as the natural sciences and Christian theology, the focus of this work.

---

[25] Whewell, *The Philosophy of the Inductive Sciences*, vol. 2, 46.

In such cases, any such colligation will be contested, not least because of the methodological asymmetry between the disciplines. Such a colligation is to be seen as characterized by two principles:

1. It is *a matter of choice*, reflecting the personal epistemic conclusions of the individual thinker (or a group of thinkers), rather than being compelled by publicly unassailable evidential warrants. As has been emphasized in this work, such a commitment will rest on what a given thinker may consider to be a rational judgement for reasons that may not be seen as compelling by those within other traditions of rationality.
2. It usually involves making connections at *different levels* of analysis, especially in colligating levels of functional explanation on the one hand, and levels of meaning or value on the other. Once more, this does not involve suspension of rational judgement on the part of individuals or communities wishing to make such connections.

To illustrate these points, we shall consider the way in which the natural sciences might be brought into conversation with socialism.

A CASE STUDY IN COLLIGATION: SCIENCE AND SOCIALISM

The analysis presented in this volume suggests that each intellectual discipline develops its own distinct research methods, embodying its own account of rationality, in a manner that is found to be appropriate for its specific disciplinary tasks and goals. The application of these methods results in certain intellectual outcomes. As we have stressed, these outcomes are often to be seen as provisional, reflecting a historically situated consensus within a given epistemic community, which is subject to revision and correction in the light of continuous theoretical reflection and the accumulation of relevant evidence. Yet a diversity of methods does not preclude a correlation of their outcomes.

A case study may be helpful in illuminating this point. Mary Midgley has noted how Marxism, one of the 'two great secular faiths of our day', exhibits 'religious-looking features',[26] thus suggesting that attempts to correlate science and socialism might serve as an analogy for the correlation of science and theology, if only to some limited extent. Although Karl Marx did not give a particularly high profile to the scientific status of socialism in his early works,[27] late nineteenth-century Marxism came to regard it as integral to its self-understanding.[28] Friedrich Engels' famous essay 'Socialisme utopique et socialisme scientifique' (1880) is generally regarded as crystallizing the core themes of such a scientific socialism.[29]

In its strong form, the notion of 'scientific socialism' articulates what one might term the 'scientific inevitability of socialism'. Perhaps the most familiar statement of this is found in Engels' address at Karl Marx's funeral at Highgate Cemetery, London in March 1883: 'just as Darwin discovered the law of development of organic nature, so Marx discovered the law of development of human history.'[30] Marx sought to formalize and codify socialism as a socio-economic theory that was rigorously grounded in the objective scientific study of history.[31] This 'scientific socialism (*wissenschaftlicher Sozialismus*)' established as its primary principle the belief—which it presented as an objective outcome of scientific research—that all historical eras are shaped by their economic conditions, and give rise to inequalities in political, social, and economic power leading to social stratification. This process was expedited by the rise of industrial capitalism during the second half of the nineteenth century. For Engels, the system of scientific socialism described the inevitable collapse of capitalism and its subsequent replacement by a classless and stateless socialist system. Few now regard this as 'scientific' in any meaningful sense of

---

[26] Midgley, *Evolution as a Religion*, 17–18. Midgley holds that the second such 'secular faith' is Darwinism.
[27] Marx's relatively few references to the natural sciences are discussed in Griese and Sandkühler, eds. *Karl Marx—Zwischen Philosophie und Naturwissenschaften*.
[28] Quante, Schweikard, and Hoesch, *Marx-Handbuch*, 280–304.
[29] A German translation appeared in 1882. For its basic themes, see Thomas, *Marxism and Scientific Socialism*, 35–49.
[30] Thomas, *Marxism and Scientific Socialism*, 1–2.
[31] Höppner, 'Karl Marx—Begründer des Wissenschaftlichen Kommunismus'. Marx's analysis echoes the strongly positivist understanding of history then emerging within Germany: Beiser, *The German Historicist Tradition*, 254–6.

the term, not least because it presupposed its core doctrine of class warfare, and retrojected this on to earlier periods of history. However, this is not the core issue for our purposes; the key point is that socialism—a political system of meaning and value—was correlated with an allegedly 'objective' scientific analysis of history and culture, which was argued to predict the inevitable triumph of those values.[32]

Yet alongside this, we must note a second approach, which is weaker in its apologetic claims yet potentially far more plausible as a meaningful worldview—namely, the notion that there is a fundamental resonance between socialist values and a scientific culture.[33] Although there were attempts to develop socialist theory in ways that paralleled the objectivity of the sciences, in order to confer on it a comparable objectivity, many German socialists of the late nineteenth century regarded the natural sciences and socialism as sharing progressivist values, allowing for an at least partial alignment of their ideologies.[34] Stuart Hampshire argued that socialism was a 'set of moral injunctions which seem to [him] clearly right and rationally justifiable';[35] yet the methods by which one might demonstrate that such values are 'rationally justifiable' will not be the same as those by which one might demonstrate that the universe is 13.8 billion years old. Yet this does not prevent a contemporary cosmologist from being a socialist. Different rational toolkits have to be developed and adapted for different tasks and goals; yet this does not mean that their outcomes cannot be correlated within a thinking person's view of the world.

So what interpretative frameworks might inform this colligation of science and socialism? How is this process of colligation to be visualized? As we have seen (59–65), the visual metaphor of perspectives is capable of incorporating both the notion of different angular or spatial perspectives on reality, and multiple levels of reality. Those who wish to defend this colligation are neither confusing nor identifying socialism with science—for example, by blurring their

---

[32] For the later development and application of these ideas, see Joravsky, *Soviet Marxism and Natural Science, 1917–1932*.

[33] It remains an open question, of course, what those values are, and the extent to which they are distinctive of socialism: see especially Collier, 'Scientific Socialism and the Question of Socialist Values'.

[34] See the careful study of Bayertz, 'Naturwissenschaft und Sozialismus'.

[35] Kołakowski and Hampshire, *The Socialist Idea*, 249.

distinctive boundaries. Rather, it is a process of weaving multiple threads together, without losing sight of the identity and distinctive nature of those threads, and remaining interested in seeing what pattern emerges from this process of aligning and interconnecting.

It might be objected that this is a fuzzy way of engaging our world, lacking the clinical precision of logic. This is a fair point; yet it needs to be put into context. It is an empirical observation that human beings seem to construct their identities and understand their worlds using multiple narratives,[36] bringing religious, political, social, and cultural ideas together within an imaginative framework as we try to make sense of things. There is no logical contradiction here, in that the insights of the natural sciences and socialist theory operate at different levels. Indeed, these could be supplemented further, with the addition of a Christian perspective, such as that found in the Christian socialism of R. H. Tawney.[37]

The point being made here is that a commitment to both the scientific method and to socialism, whether as a set of ideas or as embodied action, is not intellectually incoherent. It might be questioned by those with alternative political perspectives; yet *generically* the colligation of these two perspectives is intellectually legitimate, and is easily instanced from leading figures in the past and present.

Albert Einstein is a good example of a physicist with strongly socialist inclinations who saw his political and social values as interconnected with—though not determined by—the scientific method. In many ways, Einstein is an example of the kind of person that E. O. Wilson described as a 'synthesizer'—someone who can bring multiple disciplines together, and open up a deeper and richer way of envisioning our world, and acting within it.[38] Indeed, Einstein models the interdisciplinary 'consilience'—the term is borrowed from Whewell—that Wilson believes is essential for contemporary culture, especially through bringing the science and humanities into productive dialogue. The increasing complexity of the world mandates such

---

[36] Smith, *Moral, Believing Animals*, 63-94.
[37] Marsden, 'Richard Tawney: Moral Theology and the Social Order'.
[38] Wilson, *Consilience*, 294.

transgression of disciplinary boundaries, in order to achieve wisdom and insight. Einstein saw this as socially significant; more important, he also saw it as intellectually acceptable and meaningful.

In his neglected 1949 essay 'Why Socialism?'[39] Einstein argued that the natural sciences cannot create moral goals, even though it may provide means by which they could be achieved. Those goals do not originate from science; yet science might be the catalyst for their implementation. It is interesting to note that Einstein had used a similar intellectual framework ten years earlier in arguing for a constructive engagement between the natural sciences and religion.[40] Einstein here argued that 'the scientific method can teach us nothing else beyond how facts are related to, and conditioned by, each other'. Human beings, however, need more than what a 'purely rational conception of our existence' is able to offer. Yet opening up such fundamental questions of meaning and value does not cause us to lapse into some kind of superstitious irrationality. 'Objective knowledge provides us with powerful instruments for the achievements of certain ends, but the ultimate goal itself and the longing to reach it must come from another source.' For Einstein, the fundamental beliefs which are 'necessary and determinant for our conduct and judgments' cannot be developed or sustained in a 'solid scientific way'. Einstein thus framed the natural sciences in a way that created space for moral or spiritual values—including those arising from socialism or religion.

We need, however, to return to Stuart Hampshire to make a further point. Noting that human beings have lived on the basis of very different conceptions of the 'good life'—including socialism— Hampshire argues that no individual or community can avoid conflicts arising from incompatible moral beliefs, which cannot be resolved intellectually.[41] Philosophers may have tried in the past to find some underlying moral idea of justice which could resolve these conflicts, and would be valid at all times and in all places; these efforts, however, have been unsuccessful.[42] In one sense, therefore, Einstein's

---

[39] Einstein, *Ideas and Opinions*, 151–8.
[40] Einstein, *Ideas and Opinions*, 41–9.
[41] Hampshire, *Justice is Conflict*, 37.
[42] See especially the extended argument in Hampshire, *Innocence and Experience*.

colligation of physics and socialism involves bringing a set of robust beliefs arising from the empirical method together with a set of somewhat more tentative (in terms of their epistemic grounds) beliefs concerning morality and social justice. These two sets of beliefs are not merely derived by different methods and operative rationalities; their epistemic warrants are significantly different.

This might lead some to draw the conclusion that this colligation is unwarranted, perhaps even irrational, on the basis of the concern that these two sets of principles cannot be adhered to simultaneously without some degree of cognitive dissonance, incoherence, or inconsistency. There may well be merit in this concern; yet we have to come to terms with the brute fact that human beings do this all the time, and do not see fundamental difficulties in doing so. This colligation is psychologically plausible, and generates practices that are found to be meaningful. Natural scientists *do* hold political beliefs,[43] despite their failure to conform to the criteria of scientific rationality, and find ways of integrating these beliefs and values within their personal maps of reality. The political commitments of leading mathematicians are also instructive in making this same point.[44]

These reflections serve to reinforce the fundamental point to emerge from the analysis of rationality presented in this volume—namely, that people are perfectly capable of weaving together outcomes or insights from multiple disciplines and sources, characterized by quite distinct procedures and rationalities, without losing their character as rational reflective agents. Our complex world requires a plurality of rational methods for its investigation and meaningful inhabitation, and no one such method can be regarded as normative in every respect. Normativity is discipline-specific. We cannot develop a hybrid rationality which is valid across disciplines; rather, we have to work with the specific rationalities of each discipline, and seek to bring together their outcomes as best we can.

---

[43] See the 2009 survey conducted by the Pew Research Center for the People & the Press in collaboration with the American Association for the Advancement of Science (AAAS).
[44] Alberts, 'On Connecting Socialism and Mathematics'.

A CASE STUDY IN COLLIGATION: SCIENCE AND THEOLOGY

As our discussion of the relation of the natural sciences and socialism has made clear, it is both intellectually possible and legitimate for scientists to hold political commitments, despite the obvious rational asymmetry between scientific and political beliefs. The latter reflect beliefs and values which lie beyond the scope of the scientific method; they are not to be considered anti-scientific for that reason, but simply non-scientific. Similarly, the natural sciences and Christian theology belong at different locations on intellectual maps. It is not as if Christian theology offers a deficient or outdated version of science, which is corrected or discredited by professional scientists. It is certainly true that some early twentieth-century anthropologists took the view that religion is a primitive form of science, perhaps choosing to read the Bible as offering a deficient scientific account of the origins of the universe, when its primary concern is clearly to deal with ultimate questions about the nature and destiny of humanity, and how God relates to and transforms the human situation.[45]

Science and theology are different; there may be important boundary issues that need to be engaged, but they are not competing for the same explanatory territory, even though some representatives of each make inflated, occasionally even imperialist, claims on the other's domain.[46] Some scientists may object to religion on the ground that it appeals to the category of the 'supernatural', which is clearly at odds with scientific naturalism. Yet terms such as 'spiritual' or 'supernatural' are notoriously resistant to precise definition. Paul Draper suggests that a 'supernatural entity' is one that can affect the natural world without being part of it.[47] By definition, this includes God; yet it also seems to include Platonic mathematical realities, which we *discover* rather than *invent*—an idea that is integral to some mathematical conceptualizations of our universe and its functioning.[48] Theologians

---

[45] See Jones, *Can Science Explain Religion?*, 187–8.
[46] See Harrison, *The Territories of Science and Religion*.
[47] For an attempt at definition, see Draper, 'God, Science, and Naturalism'.
[48] For these notions and their important to scientific theorizing, see Penrose, *The Road to Reality*, 11–17, 1028–9; Ye, 'Naturalism and Abstract Entities'.

would further object that this reduces God to an entity within the universe, rather than the grounds of that universe. 'Rationalism tries to find a place for God in its picture of the world. But God... is rather the canvas on which the picture is painted, or the frame in which it is set.'[49]

In his important and influential account of the rationality of traditions (81-3), Alasdair MacIntyre suggests that a viable intellectual tradition should be able to offer a rational account of the existence of rival traditions. Can each of these epistemic communities make sense of the existence of the other, based on their own distinct rationality?[50] The intellectual resilience of both science and theology is evident in the fact that both offer, from their own perspectives, scientific explanations of religious belief and behaviour, and theological explanations of the legitimacy and scope of the natural sciences.[51] The success of such explanations may be contested within and across these two traditions; yet they both offer rational resources that allow such explanations to be developed and applied. There is, as we noted earlier (31; 48), no tradition-independent perspective on this issue, no normative and objective 'view from nowhere' which can be justified on the basis of either a privileged perspective or a universal rationality. Perhaps this view enjoyed a 'Golden Age' in Western Europe during the eighteenth century, before the critical interrogation of such ideas gained pace in the later twentieth century.

As we argued earlier (59-65), the notion of the natural sciences and Christian theology offering differing perspectives on a complex reality has considerable potential in facilitating this process of colligation of insights, especially when this is understood to combine different angles of approach and different levels of engagement. For example, there is a growing realization that meaning is rarely bundled with scientific discovery; it must be created through narratives of the kind in which Christian theology excels.

---

[49] Inge, *Faith and Its Psychology*, 197. For Kepler's appeal to theological notions in articulating the unity of nature, see Hon, 'Kepler's Revolutionary Astronomy', 162-73.

[50] It is thus no accident that Christian theologians from the second century onwards have taken care to show how a Christian metanarrative is able to account for the existence of other religious traditions, as well as presenting Christianity as their fulfilment. For some recent examples of this approach, see Sparks, 'The Fulfilment Theology of Jean Daniélou, Karl Rahner and Jacques Dupuis'. More generally, see Sandler, 'Christentum als große Erzählung'.

[51] See Jones, *Can Science Explain Religion?*, 14-97.

A particularly important instance of this issue relates to the question of human nature and identity—always an important topic, but increasingly pertinent to the growing debates about the ethics of human technological enhancement. How might scientific and theological insights be colligated? Are they simply being *aggregated*, and thus treated as disconnected and discrete insights, which are unaffected by being brought together in this? Or are they being *integrated*, by being allowed to interact, challenge, and enrich each other—and thus being changed by this process of interactive juxtaposition? Or might there be a third possibility—a conceptual *novum*, which emerges from the process of interaction of science and theology, yet is vouchsafed by neither of these in isolation?

Consider the following three aspects of human nature, drawn respectively from evolutionary biology, cultural anthropology, and Christian theology. All would command a degree of support within their respective epistemic communities, while remaining subject to continuing reflection and evaluation.

1. Human beings are genetically predisposed to act in selfish ways, even if that may occasionally express itself in seemingly altruistic modes of behaviour.[52]
2. Human beings are predisposed to use narratives to recall the past, preserve their individual and communal identity, and construct systems of meaning.[53]
3. Human beings are sinful, alienated from God and their true destiny, and stand in need of salvation.[54]

These three ideas could easily be placed in isolated compartments, treated as disconnected concepts. Yet they can also be interconnected by means of an appropriate framing device or perspective.

For example, many consider that narratives serve an explanatory function.[55] All three insights noted above can be woven together

---

[52] The position of Richard Dawkins. For a critical discussion of its origins and legacy, see McGrath, *Dawkins' God*, 32–56.

[53] Ochs and Capps, 'Narrating the Self'; Hinchman and Hinchman, eds, *Memory, Identity, Community*.

[54] For this theme in Augustine, see Couenhoven, *Stricken by Sin, Cured by Christ*, 19–57.

[55] See the discussion in Klauk, 'Is There Such a Thing as Narrative Explanation?'

through allowing the Christian narrative of sin and the scientific narrative of evolution to inform and critique each other.[56] Such an interweaving of narratives respects their distinct identities (and the manner of their generation), while aiming to encourage cross-fertilization and enrichment. It is not difficult to see how this could be extended—for example, by asking how the Christian narrative of creation interacts with the scientific narrative of the origins of the universe.

As already noted, the spatial framework of perspectives and levels provides a convenient and imaginatively plausible means of holding together the ontological unity of nature and the deployment of a plurality of methods in its exploration. 'Creation' and 'origination' can be seen as different perspectives on reality, each embedded within its own set of assumptions. The functional explanation of the universe can be enriched and expanded by recognizing the possibility of levels of meaning and value, which exist and are discerned using different conceptual toolboxes. Such an approach moves us far beyond the arid explanatory reductionism we find in some sectors of the natural sciences, opening up the discussion of human identity and functionality in productive and illuminating manners—not least by challenging the explanatory finality and totality of any single discipline, or its practical applicability to the entire range of tasks and problems facing human beings, as individuals and social communities.

Yet the object of this monograph is not to answer these questions, but to show that a rational approach can be developed to allow them to be engaged, leading to the weaving together of multiple insights and approaches. In what follows, we shall offer a brief sketch of what one such approach might look like.

RATIONALITY: A COHESIVE APPROACH

This work has offered a detailed analysis of human rationality, primarily in the natural sciences and secondarily in Christian theology, aiming to move the framework of discussion away from an outdated

---

[56] For an excellent example of such an approach, see Nielsen, *Sin and Selfish Genes*.

notion of a single rationality to which all disciplines should conform, to a range of multiply situated rationalities, which arise authentically and naturally as a result of the specificity of the various forms of human knowledge production. Its emphasis has fallen on mapping the territories of human reason, as a prelude to the task of offering an enriched account of the possible interweaving and interconnection of the natural sciences and Christian theology.

The main concern of this work is thus to provide intellectual justification for and facilitation to a meaningful conversation between the natural sciences and Christian theology, without compromising the rationality of this joint enterprise—or the two individual enterprises—in the first place. While this focus on the relation of science and theology has wider interdisciplinary relevance, it is treated here as a subject of significance in its own right, given the high cultural profiles of both science and religion, and especially the continuing importance of finding meaning for the subjective well-being of individuals.

This work has set out to map the territories of human reason. Cultural psychology suggests that the abstract potentialities of the human mind—such as the abilities to think and act, to develop norms and values, and to discern purpose and meaning—cannot be understood to exist outside a cultural context. They are, so to speak, emergent, realizing their full potential and displaying their developed forms within the behavioural and symbolic context of a community.[57] The exercise of human cognition within specific cultural and disciplinary contexts leads to the conceptual fragmentation of our single world, partly through focussing on only certain aspects of that world, and partly because of the emergence of distinct notions of rationality in response to specific disciplinary tasks. The question is whether these fragmentary insights can be reconnected, so that a cohesive account of our world can be reconstituted.

We may live in a world that is an ontological unity, but this world is investigated and represented on the basis of an epistemological pluralism, offering us a *bricolage* of unintegrated insights and perceptions arising from different disciplinary or cultural perspectives on our

---

[57] See especially Cassaniti and Menon, eds, *Universalism without Uniformity*.

world, or scientific engagement with its different levels.[58] It is not difficult to develop theoretical nets which help us visualize the interconnectedness of these perceptions, or understand how they arise from the exercise of their specific disciplinary rationalities. What remains stubbornly elusive is their failure to provide means of calibrating their respective reliabilities, or to offer a theoretical framework that allows the settling of boundary disputes, or a precise coordination of their interactions.

There are many strategies that might be developed and deployed to cope with this situation. This diversity might be seen simply as something which is to be acknowledged and respected, rather than as something which can be integrated into a coherent theoretical framework. On this approach, the important thing is to hold these insights together, without resolving the tensions that this causes. This is pragmatically helpful—for example, in allowing an enhanced appreciation of the complexity and dynamics of human communities.[59] It is, however, an unsatisfactory strategy at a theoretical level, offering apparently eclectic and descriptive accounts of a complex reality which are seen to lack conceptual rigour. While its pragmatism may seem to some to be the only realistic way forward, it lacks the capacity to offer a convincing intellectual account of the coordination of such insights and outcomes, and at best leads to these insights being placed in parallel columns, without any hope of integration or connection.

This difficulty is particularly evident in what now seems to be the failure of attempts to develop a comprehensive account of human nature. Reductionist approaches may have a certain conceptual neatness and simplicity; they fail, however, to do justice to the complexity of human nature, or to account for the emergence of certain properties or functions at higher levels of the system.[60] Perhaps even more problematically, there is disagreement about the fundamental nature

---

[58] I borrow this phrase from Nina Lykke, who speaks of the transgressing of disciplinary boundaries as allowing for 'a theoretical and methodological bricolage that allows for new synergies to emerge': Lykke, 'Women's/Gender/Feminist Studies—A Post-Disciplinary Discipline?', 96.

[59] For a good example of this kind of approach, and an exploration of its practical benefits, see Larsen and Stock, 'Capturing Contrasted Realities'.

[60] For a detailed discussion, see Ellis, *How Can Physics Underlie the Mind?*, 291–382.

of the perspectives to be integrated, and the manner in which this is to be achieved.[61] How are such disciplinary insights to be ranked in any such integrative approach? And given that so many are provisional and contested, how can a stable model result? Even a quick comparison of major recent writings on the theme shows an astonishing variety of multiperspectival scientific 'takes' on human nature, and no obvious way of resolving their divergent outcomes.[62] These concerns have led some to suggest that it is no longer meaningful to speak of 'human nature'.

A second approach is to suggest that, given the multiplicity of rationalities, it might be possible to proceed on the basis of their lowest common denominator—in effect constructing a hybrid or pragmatic rationality for the purposes of resolving disputes or incorporating different perspectives into a decision-making process.[63] Such an approach may have merits in some contexts; it remains, however, of questionable academic utility, given its pragmatic orientation. The really interesting question concerns not the baseline, but what may be built upon it.

A third approach might be to develop a 'meta-rationality'—a grand theory of human rationality which is capable of accommodating such divergences across disciplines, and allowing them to be seen within a greater whole. This is perhaps the most ambitious way of engaging the issue, and has considerable imaginative appeal. There are, however, some troubling questions here about perspectival privilege, which parallel those associated with intellectually flawed attempts to deal with religious diversity by proposing that all religious traditions are to be seen as different yet equally valid responses to 'The Real'.[64] There is a serious risk of intellectual colonization, in which the perspectives of one epistemic community are treated as normative

---

[61] See the important discussions in Fuentes and Visala, eds, *Verbs, Bones, and Brains*.

[62] Compare, for example, the approaches and outcomes found in two influential recent works: Pinker, *The Blank Slate*, and Prinz, *Beyond Human Nature*. For a helpful summary of such divergent approaches, see Pojman, *Who Are We?*

[63] Bouwmeester, *The Social Construction of Rationality*, 18–43. For an excellent example of such an approach in practical situations, see Li, Ashkanasy, and Ahlstrom, 'The Rationality of Emotions'.

[64] D'Costa, 'Whose Objectivity? Which Neutrality?'

or privileged, and in effect are used to enfold other communities within its scope.

For the Christian theologian, one attractive possibility is to see a Trinitarian 'big picture' as having the intellectual and imaginative capacity to function as such a 'meta-rationality', for example in accommodating the existence of other religious traditions, or providing a framework for grasping the relation of the sciences and theology.[65] While this 'meta-rationality' is specific to the Christian tradition, and thus carries little epistemic weight beyond its boundaries, it nevertheless informs a distinctively Christian approach to an understanding of the relation of the natural sciences and theology, and provides a robust foundation and framework for transdisciplinary conversations and research. Such rational specificity is to be seen not as a liability, but as a distinctive characteristic which plays into and informs any Christian discussion of the relation of theology and other disciplines, and—at least to a Christian constituency—offers a theoretical framework capable of holding together multiple insights concerning our complex world. This possibility clearly needs further development, which cannot be undertaken in this present volume, which has focussed rather on the question of human rationality which underlies such an enterprise.

So what might be concluded? If E. O. Wilson is right, the future of human civilization depends on 'synthesizers'—those who can make connections across disciplines. 'We are drowning in information, while starving for wisdom.'[66] As Wilson recognizes, science and religion are, and will remain, major elements of our culture; they need to talk meaningfully to each other. Wisdom, it is widely agreed, is not to be equated with knowledge; rather, it can be thought of as a 'deeper rationality' which represents the integration of insights, and the application of these to real life. It is more than simply knowing facts; it is about colligating them, and hence grasping and enacting their coherence-making and meaning-generating properties.

---

[65] See, for example, Kärkkäinen, *Trinity and Religious Pluralism*; Polkinghorne, 'Physics and Metaphysics in a Trinitarian Perspective'; Reich, 'The Doctrine of the Trinity as a Model for Structuring the Relations Between Science and Theology'.

[66] Wilson, *Consilience*, 294. For Wilson's assumption of the legitimate hegemony of the sciences in this conversation, see Segerstrale, 'Wilson and the Unification of Science'.

As Wilson hints, the debate has moved beyond whether the natural sciences, politics, and religious belief are 'compatible' with one another, as if there were some rational norm, some privileged point of adjudication, by which this question might be answered. The evidence is uncontestable: many people do consider these multiple perspectives to be mutually compatible, even enriching. The conversation needs to move on, perhaps drawing on the insights of the psychology of cognitive interdisciplinarity, which actively seeks to understand how individuals can—and *do*—hold together multiple perspectives, drawn from different disciplines.[67] How can we reverse the excesses of specialization, and create room for interdisciplinary or transdisciplinary cross-fertilization? One route is through the cultivation of epistemic dependence, in which multiple thinkers bring about a shared expansion of a group's vision; another is the rediscovery of the Renaissance vision, in which individuals try to absorb as much knowledge as they can in their quest for wisdom. There is, however, a limit to what an individual can grasp of a discipline without being an active participant in its community of research.

We need a better approach than that which Wilson himself offers to achieve the noble ambition of consilience—namely, an approach that is responsive to the massive shifts in our understanding of the territory of human reason, which require us to leave behind some of the Enlightenment's own methods and assumptions if we are to achieve some of its broader goals. Perhaps the approach set out in this work provides a more reliable and generous framework for that conversation, offering intellectual hospitality to multiple approaches, and welcoming all to the table of rational discussion. The complexity of the question and its attending issues, however, are such that this conversation may never reach a firm conclusion, however interesting and productive it may become.

---

[67] For an important study, see Bromme, 'Beyond One's Own Perspective'.

Bibliography

Abel, Günter. 'Der interne Zusammenhang von Sprache, Kommunikation, Lebenswelt und Wissenschaft.' In *Lebenswelt und Wissenschaft*, edited by Carl F. Gethmann, 351–71. Hamburg: Meiner Verlag, 2011.
Abernathy, David B. *The Dynamics of Global Dominance: European Overseas Empires, 1415–1980*. New Haven, CT: Yale University Press, 2001.
Absher, Brandon. 'Speaking of Being: Language, Speech, and Silence in *Being and Time*.' *Journal of Speculative Philosophy* 30, 2 (2016): 204–31.
Achinstein, Peter. *The Nature of Explanation*. Oxford: Oxford University Press, 1983.
Achinstein, Peter. *The Book of Evidence*. Oxford: Oxford University Press, 2001.
Achourioti, Theodora, Andrew J. B. Fugard, and Keith Stenning. 'The Empirical Study of Norms Is Just What We Are Missing.' *Frontiers in Psychology* 5 (2014), 1159.
Adam, Matthias. *Theoriebeladenheit und Objektivität. Zur Rolle von Beobachtungen in den Naturwissenschaften*. Frankfurt am Main: Ontos Verlag, 2002.
Agazzi, Evandro. *Scientific Objectivity and Its Contexts*. New York: Springer, 2014.
Akeroyd, F. Michael. 'Mechanistic Explanation Versus Deductive-Nomological Explanation.' *Foundations of Chemistry* 10, 1 (2008): 39–48.
Alberts, Gerard. 'On Connecting Socialism and Mathematics: Dirk Struik, Jan Burgers, and Jan Tinbergen.' *Historia Mathematica* 21 (1994): 280–305.
Allen, Amy. 'Macintyre's Traditionalism.' *Journal of Value Inquiry* 31 (1997): 511–25.
Alston, William P. 'A "Doxastic Practice" Approach to Epistemology.' In *Knowledge and Skepticism*, edited by Marjorie Clay and Keith Lehrer, 1–29. Boulder, CO: Westview Press, 1989.
Anderson, Philip W. 'Science: A "Dappled World" or a "Seamless Web"?' *Studies in the History and Philosophy of Modern Physics* 32, 3 (2001): 487–94.
Apel, Karl-Otto. 'The *Erklären-Verstehen* Controversy in the Philosophy of the Natural and Human Sciences.' In *Contemporary Philosophy: A New Survey*, edited by G. Floistad, 19–49. The Hague: Nijhof, 1982.
Apel, Karl-Otto and Matthias Kettner, eds. *Die eine Vernunft und die vielen Rationalitäten*. Frankfurt am Main: Suhrkamp, 1996.

Arce, Alberto and Norman Long, eds. *Anthropology, Development and Modernities: Exploring Discourses, Counter-Tendencies and Violence.* London: Routledge, 2000.

Arenas, Luis. 'Matemáticas, Método y Mathesis Universalis en las *Regulae* de Descartes.' *Revista de Filosofía* 8 (1996): 37–61.

Ariso, José María. 'Unbegründeter Glaube bei Wittgenstein und Ortega y Gasset: Ähnliche Antworten auf unterschiedliche Probleme.' *Wittgenstein-Studien* 2 (2011): 219–47.

Arnason, Johann P. 'The Multiplication of Modernity.' In *Identity, Culture, and Globalization*, edited by Eliezer Ben Rafael and Yitzak Sternberg, 131–54. Leiden: Brill, 2001.

Atkins, Peter. 'The Limitless Power of Science.' In *Nature's Imagination: The Frontiers of Scientific Vision*, edited by John Cornwell, 122–32. Oxford: Oxford University Press, 1995.

Atkins, Peter. *On Being: A Scientist's Exploration of the Great Questions of Existence.* Oxford: Oxford University Press, 2011.

Audi, Robert. 'Theoretical Rationality: Its Sources, Structure, and Scope.' In *The Oxford Handbook of Rationality*, edited by Alfred R. Mele and Piers Rawling, 17–44. Oxford: Oxford University Press, 2004.

Avgerou, Chrisanthi. *Information Systems and Global Diversity.* Oxford: Oxford University Press, 2002.

Avrahami, Yael. *The Senses of Scripture: Sensory Perception in the Hebrew Bible.* London: T & T Clark, 2012.

Ayala, Francisco J. 'In William Paley's Shadow: Darwin's Explanation of Design.' *Ludus Vitalis* 12, 21 (2004): 53–66.

Ayala, Francisco J. 'Darwin and the Scientific Method.' *Proceedings of the National Academy of Sciences* 106, Supplement 1 (2009): 10033–9.

Bacon, Francis. *The New Organon.* Cambridge: Cambridge University Press, 2000.

Baehr, Jason S. *The Inquiring Mind: On Intellectual Virtues and Virtue Epistemology.* Oxford: Oxford University Press, 2012.

Baehr, Peter. 'The"Iron Cage" and the "Shell as Hard as Steel:" Parsons, Weber, and the Stahlhartes Gehäuse Metaphor in the Protestant Ethic and the Spirit of Capitalism.' *History and Theory* 40, 2 (2001): 153–69.

Baker, Alan. 'Occam's Razor in Science: A Case Study from Biogeography.' *Biology and Philosophy* 22 (2007): 193–215.

Baker, Gordon P. *Wittgenstein's Method: Neglected Aspects. Essays on Wittgenstein.* Malden, MA: Blackwell, 2004.

Ball, Stephen W. 'Gibbard's Evolutionary Theory of Rationality and Its Ethical Implications.' *Biology and Philosophy* 10, 1 (1995): 129–80.

Bamford, Greg. 'Popper and His Commentators on the Discovery of Neptune: A Close Shave for the Law of Gravitation?' *Studies in History and Philosophy of Science Part A* 27, 2 (1996): 207–32.
Bangu, Sorin. 'Scientific Explanation and Understanding: Unificationism Reconsidered.' *European Journal for Philosophy of Science* 7, 1 (2017): 103–26.
Banner, Michael C. *The Justification of Science and the Rationality of Religious Belief.* Oxford: Oxford University Press, 1990.
Barnes, Eric C. 'Inference to the Loveliest Explanation.' *Synthese* 103 (1995): 251–78.
Barnes, Eric C. *The Paradox of Predictivism.* Cambridge: Cambridge University Press, 2008.
Barrett, Justin L. *Why Would Anyone Believe in God?* Lanham, MD: AltaMira Press, 2004.
Barrett, Justin L. *Born Believers: The Science of Children's Religious Belief.* New York: Free Press, 2012.
Barros, Gustavo. 'Herbert A. Simon and the Concept of Rationality: Boundaries and Procedures.' *Brazilian Journal of Political Economy* 30, 3 (2010): 455–72.
Barrow, John D. 'Mathematical Explanation.' In *Explanations: Styles of Explanation in Science*, edited by John Cornwell, 81–109. Oxford: Oxford University Press, 2004.
Bartelborth, Thomas. 'Explanatory Unification.' *Synthese* 130 (2002): 91–108.
Bates, David William. *Enlightenment Aberrations: Error and Revolution in France.* Ithaca, NY: Cornell University Press, 2002.
Bauckham, Richard. *Jesus and the God of Israel: God Crucified and Other Studies on the New Testament's Christology of Divine Identity.* Grand Rapids, MI: Eerdmans, 2008.
Bauman, Zygmunt. *Legislators and Interpreters: On Modernity, Post-Modernity, and Intellectuals.* Cambridge: Polity Press, 1987.
Baxter, Richard. The *Practical Works of Richard Baxter.* 23 vols. London: James Duncan, 1830.
Bayer, Oswald. *Vernunft ist Sprache: Hamanns Metakritik Kants.* Stuttgart: Frommann-Holzboog Verlag, 2002.
Bayertz, Kurt. 'Naturwissenschaft und Sozialismus: Tendenzen der Naturwissenschafts-Rezeption in der Deutschen Arbeiterbewegung des 19. Jahrhunderts.' *Social Studies of Science* 13, 3 (1983): 323–53.
Beall, J. C. and Greg Restall. *Logical Pluralism.* Oxford: Clarendon Press, 2006.
Beam, Craig Allen. 'Gadamer and MacIntyre: Tradition as a Resource of Rationality.' *Kinesis* 25 (1998): 15–35.

Bechtel, William. 'Levels of Description and Explanation in Cognitive Science.' *Minds and Machines* 4, 1 (1994): 1–25.
Bechtel, William. 'Explicating Top-Down Causation Using Networks and Dynamics.' *Philosophy of Science* 84 (2017): 253–74.
Bechtle, Gerald. 'How to Apply the Modern Concepts of *Mathesis Universalis* and *Scientia Universalis* to Ancient Philosophy: Aristotle, Platonisms, Gilbert of Poitiers, and Descartes.' In *Platonisms: Ancient, Modern, and Postmodern*, edited by Kevin Corrigan and John D. Turner, 129–54. Leiden: Brill, 2007.
Beck, Ulrich, Wolfgang Bonss, and Christoph Lau. 'The Theory of Reflexive Modernization: Problematic, Hypotheses and Research Programme.' *Theory, Culture & Society* 20, 2 (2003): 1–33.
Beiser, Frederick C. *The Sovereignty of Reason: The Defense of Rationality in the Early English Enlightenment*. Princeton, NJ: Princeton University Press, 1996.
Beiser, Frederick C. *The German Historicist Tradition*. Oxford: Oxford University Press, 2015.
Berger, Peter L. *A Far Glory: The Quest for Faith in an Age of Credulity*. New York: Free Press, 1992.
Bernard, Claude. *Introduction à l'étude de la médecine expérimentale*. Paris: Ballière, 1865.
Bertolotti, Tommaso, ed. *Patterns of Rationality: Recurring Inferences in Science, Social Cognition and Religious Thinking*. Heidelberg: Springer, 2015.
Beveridge, W. I. B. *The Art of Scientific Investigation*. New York: Norton, 1957.
Bhaskar, Roy. *The Possibility of Naturalism: A Philosophical Critique of the Contemporary Human Sciences*. 3rd edn. London: Routledge, 1998.
Biagioli, Mario. 'Postdisciplinary Liaisons: Science Studies and the Humanities.' *Critical Inquiry* 35, 4 (2009): 816–33.
Bittner, Thomas and Barry Smith. 'A Theory of Granular Partitions.' In *Applied Ontology: An Introduction*, edited by Katherine Munn and Barry Smith, 125–58. Frankfurt: Ontos, 2008.
Blachowicz, James. 'How Science Textbooks Treat Scientific Method: A Philosopher's Perspective.' *British Journal for the Philosophy of Science*, 60, 2 (2009): 303–44.
Boff, Leonardo. *Trinity and Society*. London: Burns & Oates, 1988.
Boghossian, Paul. *Fear of Knowledge: Against Relativism and Constructivism*. Oxford: Oxford University Press, 2006.
Bolton, Robert. 'The Epistemological Basis of Aristotle's Dialectic.' In *From Puzzles to Principles: Essays on Aristotle's Dialectic*, edited by May Sim, 57–106. Lanham, MD: Lexington Books, 1999.

Bonk, Thomas. *Underdetermination: An Essay on Evidence and the Limits of Natural Knowledge*. Boston Studies in the Philosophy of Science. Dordrecht: Springer, 2008.

Boudon, Raymond. 'The Cognitive Approach to Morality.' In *Handbook of the Sociology of Morality*, edited by Steven Hitlin and Stephen Vaisey, 15–33. New York: Springer, 2010.

Boulter, Stephen. *The Rediscovery of Common Sense Philosophy*. Basingstoke: Palgrave Macmillan, 2007.

Bouwmeester, Onno. *The Social Construction of Rationality: Policy Debates and the Power of Good Reasons*. New York: Routledge, 2017.

Boyer, Pascal. *Religion Explained. The Evolutionary Origins of Religious Thought*. New York: Basic Books, 2001.

Boyle, Deborah. 'Hume on Animal Reason.' *Hume Studies* 29, 1 (2003): 3–28.

Bradley, David F., Julie J. Exline, and Alex Uzdavines. 'The God of Nonbelievers: Characteristics of a Hypothetical God.' *Science, Religion and Culture* 2, 3 (2015): 120–30.

Brogaard, Berit and Barry Smith, eds. *Rationalität und Irrationalität: Akten des 23. Internationalen Wittgenstein-Symposiums*. Vienna: Öbv & Hpt, 2001.

Bromberger, Sylvain. 'Why-Questions.' In *Readings in the Philosophy of Science*, edited by Baruch A. Brody, 66–84. Englewood Cliffs, NJ: Prentice Hall, 1966.

Bromme, Rainer. 'Beyond One's Own Perspective: The Psychology of Cognitive Interdisciplinarity.' In *Practising Interdisciplinarity*, edited by Nico Stehr and Peter Weingart, 115–33. Toronto: University of Toronto Press, 2000.

Brooke, John Hedley. 'Like Minds: The God of Hugh Miller.' In *Hugh Miller and the Controversies of Victorian Science*, edited by Michael Shortland, 171–86. Oxford: Clarendon Press, 1996.

Brooke, John Hedley. *Science and Religion: Some Historical Perspectives*. Cambridge: Cambridge University Press, 2014.

Brooke, John Hedley. 'Living with Theology and Science: From Past to Present.' *Theology and Science* 12, 4 (2014): 307–23.

Broome, John. 'Does Rationality Give Us Reasons?' *Philosophical Issues* 15 (2005): 321–37.

Brown, Harvey R. and David Wallace, 'Solving the Measurement Problem: De Broglie-Bohm Loses Out to Everett.' *Foundations of Physics*, 35, 4 (2005): 517–40.

Browne, Craig. 'Postmodernism, Ideology and Rationality.' *Revue internationale de philosophie* 251, 1 (2010): 79–99.

Brukner, Časlav and Anton Zeilinger. 'Quantum Physics as a Science of Information.' In *Quo Vadis Quantum Mechanics? Possible Developments* of Quantum Mechanics in the 21st Century, edited by Avshalom C. Elitzur, Shahar Dolev, and Nancy Kolenda, 47–61. Berlin: Springer, 2006.

Brümmer, Vincent. 'The Intersubjectivity of Criteria in Theology.' In *Brümmer on Meaning and the Christian Faith: Collected Writings of Vincent Brümmer*, 453–70. Basingstoke: Ashgate, 2006.

Brunner, Emil. *Dogmatik I: Die christliche Lehre von Gott*. Zurich: Zwingli-Verlag, 1959.

Bühler, Pierre. 'Tertullian: The Teacher of the *Credo Quia Absurdum*.' In *Kierkegaard and the Patristic and Medieval Traditions*, edited by Jon Stewart, 131–8. Burlington, VT: Ashgate, 2008.

Bunge, Mario. 'The Weight of Simplicity in the Construction and Assaying of Scientific Theories.' *Philosophy of Science* 28, 2 (1961): 120–49.

Bunge, Mario. *Philosophy of Science: From Problem to Theory*. London: Routledge, 2017.

Burdett, Michael. 'The Image of God and Human Uniqueness: Challenges from the Biological and Information Sciences.' *Expository Times* 127, 1 (2015): 3–10.

Burns, Robert M. 'Richard Swinburne on Simplicity in Natural Science.' *Heythrop Journal* 40 (1999): 184–206.

Burton, Simon J. G. *The Hallowing of Logic: The Trinitarian Method of Richard Baxter's Methodus Theologiae*. Leiden: Brill, 2012.

Buzzoni, Marco. 'Erkenntnistheoretische und ontologische Probleme der theoretischen Begriffe.' *Zeitschrift für Allgemeine Wissenschaftstheorie* 28, 1 (1997): 19–53.

Cahan, David, ed. *From Natural Philosophy to the Sciences: Writing the History of Nineteenth-Century Science*. Chicago, IL: University of Chicago Press, 2003.

Campos, Daniel. 'On the Distinction Between Peirce's Abduction and Lipton's Inference to the Best Explanation.' *Synthese* 180 (2011): 419–42.

Cannon, Walter F. 'The Impact of Uniformitarianism: Two Letters from John Herschel to Charles Lyell, 1836–1837.' *Proceedings of the American Philosophical Society* 105 (1961): 301–14.

Cantor, Geoffrey and Chris Kenny. 'Barbour's Fourfold Way: Problems with His Taxonomy of Science-Religion Relationships.' *Zygon* 36 (2001): 765–81.

Carey, Daniel. *Locke, Shaftesbury, and Hutcheson: Contesting Diversity in the Enlightenment and Beyond*. Cambridge: Cambridge University Press, 2006.

Carpendale, Jeremy I. M. and Ulrich Miller. 'Social Interaction and the Development of Rationality and Morality: An Introduction.' In *Social Interaction and the Development of Knowledge*, edited by J. I. M. Carpendale and U. Miller, 1–18. Mahwah, NJ: Lawrence Erlbaum Associates, 2004.

Carr, Bernard, ed. *Universe or Multiverse?* Cambridge: Cambridge University Press, 2007.

Carrier, Martin. 'Values and Objectivity in Science: Value-Ladenness, Pluralism and the Epistemic Attitude.' *Science & Education* 22 (2013): 2547–68.
Carroll, Anthony J. 'Disenchantment, Rationality, and the Modernity of Max Weber.' *Forum Philosophicum* 16, 1 (2011): 117–37.
Cartwright, Nancy. *The Dappled World: A Study of the Boundaries of Science.* Cambridge: Cambridge University Press, 1999.
Cassaniti, Julia and Usha Menon, eds. *Universalism without Uniformity: Explorations in Mind and Culture.* Chicago, IL: University of Chicago Press, 2017.
Castoriadis, Cornelius. *L'institution imaginaire de la société.* Paris: Seuil, 1975.
Chang, Ruth. 'Raz on Reasons, Reason, and Rationality: On Raz's *From Normativity to Responsibility*.' *Jerusalem Review of Legal Studies* 8, 1 (2013): 1–21.
Chapman, Alister, John Coffey, and Brad S. Gregory, eds. *Seeing Things Their Way: Intellectual History and the Return of Religion.* Notre Dame, IN: University of Notre Dame Press, 2009.
Chater, Nick. 'The Search for Simplicity: A Fundamental Cognitive Principle?' *Quarterly Journal of Experimental Psychology* 52A, 2 (1999): 273–302.
Chater, Nick and Mike Oaksford. 'Normative Systems: Logic, Probability, and Rational Choice.' In *The Oxford Handbook of Thinking and Reasoning*, edited by Keith J. Holyoak and Robert G. Morrison, 11–21. Oxford: Oxford University Press, 2012.
Chen, Wein. 'Clinical Applications of PET in Brain Tumors.' *Journal of Nuclear Medicine* 48, 9 (2007): 1468–81.
Cheng-Guajardo, Luis. 'The Normative Requirement of Means-End Rationality and Modest Bootstrapping.' *Ethical Theory and Moral Practice*, 17, 3 (2014): 487–503.
Chesterton, G. K. 'The Return of the Angels.' *Daily News*, 14 March 1903.
Christianson, Gale E. *Isaac Newton and the Scientific Revolution.* New York: Oxford University Press, 1996.
Chung, Paul S. *Postcolonial Public Theology: Faith, Scientific Rationality, and Prophetic Dialogue.* Eugene, OR: Cascade Books, 2016.
Churchman, C. West. 'Science and Decision-Making.' *Philosophy of Science* 22 (1956): 247–9.
Clanton, J. Caleb. 'The Structure of C. S. Peirce's Neglected Argument for the Reality of God: A Critical Assessment.' *Transactions of The Charles S. Peirce Society* 50, 2 (2014): 175–200.
Clark, Andy. *Being There: Putting Brain, Body, and World Together Again.* Cambridge, MA: MIT Press, 1997.

Clarke, Steve and Adrian Walsh. 'Scientific Imperialism and the Proper Relations between the Sciences.' *International Studies in the Philosophy of Science* 23, 2 (2009), 195–207.

Clayton, Philip. *Explanation from Physics to Theology: An Essay in Rationality and Religion.* New Haven, CT: Yale University Press, 1989.

Clayton, Philip. 'Inference to the Best Explanation.' *Zygon* 32, 3 (1997): 177–91.

Clayton, Philip and Steven Knapp. 'Rationality and Christian Self-Conception.' In *Religion and Science: History, Method, Dialogue*, edited by W. Mark Richardson and Wesley J. Wildman, 131–44. London: Routledge, 1996.

Clerck, Paul de: see De Clerck.

Coakley, Sarah. 'Living into the Mystery of the Holy Trinity: Trinity, Prayer, and Sexuality.' *Anglican Theological Review* 80 (1998): 223–32.

Code, Murray. 'On the Poverty of Scientism, or: The Ineluctable Roughness of Rationality.' *Metaphilosophy* 28 (1997): 102–22.

Coleman, Thomas J., Ralph W. Hood, and John R. Shook. 'An Introduction to Atheism, Secularity, and Science.' *Science, Religion & Culture* 2, 3 (2015): 1–14.

Collicutt, Joanna. 'Bringing the Academic Discipline of Psychology to Bear on the Study of the Bible.' *Journal of Theological Studies* 63, 1 (2012): 1–48.

Collier, Andrew. 'Scientific Socialism and the Question of Socialist Values.' *Canadian Journal of Philosophy, Supplementary Volume* 7 (1981): 121–54.

Collier, Andrew. *Critical Realism: An Introduction to Roy Bhaskar's Philosophy.* London: Verso, 1994.

Collingwood, R. G. *An Autobiography.* London: Oxford University Press, 1939.

Collins, Harold M., Robert J. Evans, and Mike Gorman. 'Trading Zones and Interactional Expertise.' *Studies in History and Philosophy of Science Part A* 38, 4 (2007): 657–66.

Collins, Paul M. *Trinitarian Theology, West and East: Karl Barth, the Cappadocian Fathers and John Zizioulas.* Oxford: Oxford University Press, 2001.

Columbetti, Giovanna. 'Enaction, Sense-Making and Emotion.' In *Enaction: Towards a New Paradigm for Cognitive Science*, edited by J. Stewart, O. Gapenne, and E. Di Paolo, 145–64. Cambridge, MA: MIT Press, 2010.

Coman, Iacob. 'Suffering in the Trinitarian Pattern of Redemption.' *International Journal of Orthodox Theology* 4, 3 (2013): 99–127.

Conway Morris, Simon. *Life's Solution: Inevitable Humans in a Lonely Universe.* Cambridge: Cambridge University Press, 2005.

Cooper, David E. 'Wittgenstein, Heidegger, and Humility.' *Philosophy* 72 (1997): 105–23.

Cooper, David E. *The Measure of Things: Humanism, Humility and Mystery.* Oxford: Clarendon Press, 2002.
Cooper, David E. 'Living with Mystery: Virtue, Truth, and Practice.' *European Journal for Philosophy of Religion* 4, 3 (2012): 1–13.
Copan, Paul and William Lane Craig. *The Kalām Cosmological Argument: Scientific Evidence for the Beginning of the Universe.* New York: Bloomsbury Academic, 2018.
Cope, Kevin Lee. *Criteria of Certainty: Truth and Judgment in the English Enlightenment.* Lexington, KY: University Press of Kentucky, 1990.
Corsi, Pietro. *Evolution before Darwin.* Oxford: Oxford University Press, 2010.
Cottingham, John. *The Spiritual Dimension: Religion, Value and Human Life.* Cambridge: Cambridge University Press, 2005.
Cottingham, John. 'Religion and the Mystery of Existence.' *European Journal for Philosophy of Religion* 4, 3 (2012): 15–31.
Coudert, Allison. *Religion, Magic, and Science in Early Modern Europe and America.* Santa Barbara, CA: Praeger, 2011.
Couenhoven, Jesse. *Stricken by Sin, Cured by Christ: Agency, Necessity, and Culpability in Augustinian Theology.* New York: Oxford University Press, 2013.
Coulson, Charles A. *Christianity in an Age of Science.* London: Oxford University Press, 1953.
Coulson, Charles A. *Science and Christian Belief.* London: Oxford University Press, 1955.
Coulson, Charles A. *Science and the Idea of God.* Cambridge: Cambridge University Press, 1958.
Cox, Laurence and Alf Gunvald Nilsen. 'What Would a Marxist Theory of Local Movements Look Like?' In *Marxism and Social Movements*, edited by Colin Barker, 63–81. Leiden: Brill, 2013.
Craig, William Lane. *The Kalām Cosmological Argument.* London: Macmillan, 1979.
Crane, Tim. *The Meaning of Belief: Religion from an Atheist's Point of View.* Cambridge, MA: Harvard University Press, 2017.
Craver, Carl F. 'The Ontic Account of Scientific Explanation.' In *Explanation in the Special Sciences: The Case of Biology and History*, edited by M. I. Kaiser, O. R. Scholz, D. Plenge, and A. Hüttemann, 27–52. Dordrecht: Springer, 2014.
Craver, Carl F. and Marie I. Kaiser. 'Mechanisms and Laws: Clarifying the Debate.' In *Mechanism and Causality in Biology and Economics*, edited by H.-K. Chao, S.-T. Cheng, and R. L. Millstein, 125–45. New York: Springer, 2013.
Crick, Francis H. C. *The Astonishing Hypothesis: The Scientific Search for the Soul.* London: Simon & Schuster, 1994.

Cunningham, Andrew and Perry Williams. 'De-Centring the "Big Picture": The Origins of Modern Science and the Modern Origins of Science.' *British Journal for the History of Science* 26, 4 (1993): 407-32.

Curran, Andrew. *The Anatomy of Blackness: Science & Slavery in an Age of Enlightenment.* Baltimore, MD: Johns Hopkins University Press, 2011.

D'Alembert, Jean-Baptiste le Rond. *Oeuvres.* 5 vols. Paris: Belin & Bossange, 1821.

Danielson, Peter. 'Rationality and Evolution.' In *The Oxford Handbook of Rationality*, edited by Alfred R. Mele and Piers Rawling, 417-37. Oxford: Oxford University Press, 2004.

Darden, Lindley and Nancy Maull. 'Interfield Theories.' *Philosophy of Science* 44 (1977): 43-64.

Darwin, Charles. *On the Origin of the Species by Means of Natural Selection.* London: John Murray, 1859.

Darwin, Charles. *The Life and Letters of Charles Darwin*, ed. F. Darwin. 3 vols. London: John Murray, 1887.

Daston, Lorraine J. 'Objectivity and the Escape from Perspective.' *Social Studies of Science* 44, 2 (1992): 597-618.

Daston, Lorraine J. 'Scientific Error and the Ethos of Belief.' *Social Research* 72, 1 (2005): 1-28.

Daston, Lorraine and Peter Galison. *Objectivity.* New York: Zone Books, 2010.

Davies, Brian. *Thinking about God.* Eugene, OR: Wipf & Stock, 2011.

Davies, E. Brian. *Why Beliefs Matter: Reflections on the Nature of Science.* Oxford: Oxford University Press, 2014.

Davies, David. 'Living in the "Space of Reasons": The "Rationality Debate" Revisited.' *International Studies in the Philosophy of Science* 13, 3 (1999): 231-44.

Davis, Bernard D. 'The Importance of Human Individuality for Sociobiology.' *Zygon* 15, 3 (1980): 275-93.

Davis, Stephen T. *God, Reason and Theistic Proofs.* Grand Rapids, MI: Eerdmans, 1997.

Davis, Stephen T., Daniel Kendall, and Gerald O'Collins, eds. *The Trinity: An Interdisciplinary Symposium on the Trinity.* Oxford: Oxford University Press, 2002.

Dawes, Gregory W. *Theism and Explanation.* New York: Routledge, 2014.

Dawkins, Richard. *A Devil's Chaplain: Selected Essays.* London: Weidenfield & Nicolson, 2003.

Dawkins, Richard. *The God Delusion.* London: Bantam, 2006.

Dawson, Neal V. and Fredrick Gregory. 'Correspondence and Coherence in Science: A Brief Historical Perspective.' *Judgment and Decision Making* 4, 2 (2009): 126-33.

D'Costa, Gavin. 'Whose Objectivity? Which Neutrality? The Doomed Quest for a Neutral Vantage Point from Which to Judge Religions.' *Religious Studies* 29 (1993): 79-95.

Dear, Peter R. *The Intelligibility of Nature: How Science Makes Sense of the World*. Chicago, IL: University of Chicago Press, 2006.

Dear, Peter R. 'Reason and Common Culture in Early Modern Natural Philosophy: Variations on an Epistemic Theme.' In *Conflicting Values of Inquiry: Ideologies of Epistemology in Early Modern Europe*, edited by Tamás Demeter, Kathryn Murphy, and Claus Zittel, 10-38. Leiden: Brill, 2014.

De Clerck, Paul. '*Lex Orandi—Lex Credendi*: The Original Sense and Historical Avatars of an Equivocal Adage.' *Studia Liturgica* 24 (1994): 178-200.

Defez, Antoni. 'Ortega y Wittgenstein: no tan lejos.' *Revista de Filosofía* 39, 2 (2014): 81-100

Dennett, Daniel C. *Breaking the Spell: Religion as a Natural Phenomenon*. New York: Viking Penguin, 2006.

De Regt, Henk W. and Dennis Dieks. 'A Contextual Approach to Scientific Understanding.' *Synthese* 144, 1 (2005): 137-70.

Descartes, René. *Discours de la méthode*. Paris: Éditions Hatier, 1999.

De Sousa, Ronald. *Why Think? Evolution and the Rational Mind*. Oxford: Oxford University Press, 2011.

Destutt de Tracy, Antoine Louis C. *Éléments d'idéologie*. Paris: Courcier, 1801-15.

Detjen, Hans-Jürgen. *Geltungsbegründung traditionsabhängiger Weltdeutungen im Dilemma: Theologie, Philosophie, Wissenschaftstheorie und Konstruktivismus*. Berlin/Münster: LIT Verlag, 2010.

Dewey, John. *The Quest for Certainty: A Study of the Relation of Knowledge and Action*. Gifford Lectures. London: Allen & Unwin, 1930.

Dietrich, Eric. *Excellent Beauty: The Naturalness of Religion and the Unnaturalness of the World*. New York: Columbia University Press, 2015.

Díez, Jose. 'Falsificationism and the Structure of Theories: The Popper-Kuhn Controversy about the Rationality of Normal Science.' *Studies in History and Philosophy of Science* 38 (2007): 543-54.

Dijksterhuis, E. J. *De Mechanisering van het Wereldbeeld*. Amsterdam: Meulenhoff, 1996.

Dingel, Irene and Kestutis Daugirdas. *Antitrinitarische Streitigkeiten: Die tritheistische Phase (1560-1568)*. Göttingen: Vandenhoeck & Ruprecht, 2013.

Dodds, E. R. *The Greeks and the Irrational*. Berkeley, CA: University of California Press, 1951.

Douglas, Heather. 'Inductive Risk and Values.' *Philosophy of Science* 67, 4 (2000): 559-79.

Dowe, Phil. *Physical Causation*. Cambridge: Cambridge University Press, 2000.
Draper, Paul. 'God, Science, and Naturalism.' In *The Oxford Handbook of Philosophy of Religion*, edited by William J. Wainwright, 273–302. Oxford: Oxford University Press, 2005.
Ducheyne, Steffen. 'Newton on Action at a Distance.' *Journal of the History of Philosophy* 52, 4 (2014): 675–701.
Duhem, Pierre. *Sauver les apparences: Essai sur la notion de théorie physique de Platon à Galilée*. Paris: Librarie Vrin, 1908.
Duhem, Pierre. *La théorie physique: Son objet, sa structure*. Paris: Vrin, 1997 [1906].
Dukas, Helen. *Albert Einstein—The Human Side. New Glimpses from His Archives*. Princeton, NJ: Princeton University Press, 1979.
Dulles, Avery. *A History of Apologetics*. 2nd edn. San Francisco, CA: Ignatius Press, 2005.
Dunbar, Kevin N. 'How Scientists Really Reason: Scientific Reasoning in Real-World Laboratories.' In *Mechanisms of Insight*, edited by Robert J. Sternberg and Janet E. Davidson, 365–95. Cambridge, MA: MIT Press, 1995.
Dunbar, Kevin N. and David Klahr. 'Scientific Thinking and Reasoning.' In *The Oxford Handbook of Thinking and Reasoning*, edited by Keith J. Holyoak and Robert J. Morrison, 701–18. Oxford: Oxford University Press, 2012.
Dupré, John, *Human Nature and the Limits of Science*. Oxford: Clarendon Press, 2001.
Dupré, John, 'The Lure of the Simplistic.' *Philosophy of Science* 69, 3 (2002): S284–93.
Eagle, Antony. 'Randomness Is Unpredictability.' *British Journal for the Philosophy of Science* 56, 4 (2005): 749–90.
Eagleton, Terry. *Reason, Faith, and Revolution: Reflections on the God Debate*. New Haven, CT: Yale University Press, 2009.
Ecklund, Elaine Howard, David R. Johnson, Christopher P. Scheitle, Kirstin R. W. Matthew, and Steven W. Lewis. 'Religion among Scientists in International Context: A New Study of Scientists in Eight Regions.' *Socius* 2 (2016): 1–9.
Edelstein, Dan. *The Enlightenment: A Genealogy*. Chicago, IL: University of Chicago Press, 2010.
Edgerton, Samuel Y. *The Mirror, the Window and the Telescope: How Renaissance Linear Perspective Changed Our Vision of the Universe*. Ithaca, NY: Cornell University Press, 2009.
Edwards, M. J. *Image, Word and God in the Early Christian Centuries*. Farnham: Ashgate, 2013.

Egan, David. 'Pictures in Wittgenstein's Later Philosophy.' *Philosophical Investigations* 34, 1 (2011): 55–76.
Einstein, Albert. *Ideas and Opinions*. New York: Crown Publishers, 1954.
Einstein, Albert. *Mein Weltbild*, edited by Carl Seelig. Berlin: Ullstein 2005.
Eisenstadt, Shmuel N. 'Multiple Modernities.' *Daedalus* 129, 1 (2000): 1–9.
Ejsing, Anette. *Theology of Anticipation: A Constructive Study of C. S. Peirce.* Eugene, OR: Pickwick Publications, 2007.
Elgin, Catherine Z. 'Creation as Reconfiguration: Art in the Advancement of Science.' *International Studies in the Philosophy of Science* 16, 1 (2002): 13–25.
Eliade, Mircea. 'The Quest for the "Origins" of Religion.' *History of Religions* 4, 1 (1964): 154–69.
Elio, Renée, ed. *Common Sense, Reasoning, and Rationality*. New York: Oxford University Press, 2002.
Elkana, Yehuda. 'The Myth of Simplicity.' In *Albert Einstein: Historical and Cultural Perspectives*, edited by Gerald Holton and Yehuda Elkana, 205–51. Princeton, NJ: 1982.
Ellis, George. *How Can Physics Underlie the Mind? Top-Down Causation in the Human Context*. Heidelberg: Springer, 2016.
Elster, Jon. *Ulysses and the Sirens: Studies in Rationality and Irrationality*. Cambridge: Cambridge University Press, 1984.
Elster, Jon. *Ulysses Unbound: Studies in Rationality, Precommitment, and Constraints*. Cambridge: Cambridge University Press, 2000.
Elster, Jon. *Sour Grapes: Studies in the Subversion of Rationality*. New York: Cambridge University Press, 2016.
Emden, Christian J. *Nietzsche's Naturalism: Philosophy and the Life Sciences in the Nineteenth Century*. Cambridge: Cambridge University Press, 2014.
Emmons, Robert A. *The Psychology of Ultimate Concerns: Motivation and Spirituality in Personality*. New York: Guilford Press, 1999.
Engler, Gideon. 'Einstein and the Most Beautiful Theories in Physics.' *International Studies in the Philosophy of Science* 16, 1 (2002): 27–37.
Erickson, Paul, Judy L. Klein, Lorraine Daston, Rebecca M. Lemov, Thomas Sturm, and Michael D. Gordin. *How Reason Almost Lost Its Mind: The Strange Career of Cold War Rationality*. Chicago, IL: University of Chicago Press, 2013.
Esbjörn-Hargens, Sean. 'An Overview of Integral Theory: An All-Inclusive Framework for the Twenty-First Century.' In *Integral Theory in Action*, edited by Sean Esbjörn-Hargens, 33–61. Albany, NY: SUNY Press, 2010.
Euben, Roxanne L. *Fundamentalism and the Limits of Modern Rationalism: A Work of Comparative Political Theory*. Princeton, NJ: Princeton University Press, 1999.

Evans, C. Stephen. *Why Believe? Reason and Mystery as Pointers to God.* Grand Rapids, MI: Eerdmans, 1996.
Falardeau, Jean-Charles. 'Le sens du merveilleux.' In *Le merveilleux: Deuxième colloque sur les religions populaires*, edited by Fernand Dumont, Jean-Paul Montminy, and Michel Stein, 143–56. Québec: Presses de l'Université Laval, 1973.
Farrer, Austin. *The Glass of Vision.* London: Dacre Press, 1948.
Fazekas, Peter and Gergely Kertész. 'Causation at Different Levels: Tracking the Commitments of Mechanistic Explanations.' *Biology and Philosophy* 26 (2011): 365–83.
Feingold, Lawrence. *The Natural Desire to See God According to St. Thomas and His Interpreters.* Rome: Apollinare Studi, 2001.
Feist, Gregory J. *The Psychology of Science and the Origins of the Scientific Mind.* New Haven, CT: Yale University Press, 2006.
Feser, Edward. *Scholastic Metaphysics: A Contemporary Introduction.* Heusenstamm: Editiones Scholasticae, 2014.
Feyerabend, Paul K. 'Der Begriff der Verständlichkeit in der modernen Physik (1948).' *Studies in History and Philosophy of Science Part A* 57, 1 (2016): 67–9.
Feynman, Richard P. *The Character of Physical Law.* Boston, MA: MIT Press, 1988.
Fiedrowicz, Michael. *Handbuch der Patristik: Quellentexte zur Theologie der Kirchenväter.* Freiburg im Breisgau: Herder, 2010.
Filippini, Michele. *Using Gramsci: A New Approach.* London: Pluto Press, 2017.
Fischer, Ernst Peter. *Wie der Mensch seine Welt neu erschaffen hat.* Berlin: Springer Spektrum, 2013.
Fischer, Ernst Peter. *Werner Heisenberg: Ein Wanderer zwischen zwei Welten.* Berlin: Springer Spektrum, 2015.
Fitzgerald, Timothy. 'A Critique of Religion as a Cross-Cultural Category.' *Method and Theory in the Study of Religion* 9, 2 (1997): 91–110.
Fitzpatrick, Simon. 'Kelly on Ockham's Razor and Truth-Finding Efficiency.' *Philosophy of Science* 80, 2 (2013): 298–309.
Flack, Jessica C. 'Coarse-graining as a Downward Causation Mechanism.' *Philosophical Transactions of the Royal Society A* 375 (2017): 20160338.
Flyvbjerg, Bent. *Rationality and Power: Democracy in Practice.* Chicago, IL: University of Chicago Press, 1998.
Foucault, Michel. *Folie et déraison: Histoire de la folie à l'âge classique.* Paris: Union générale d'éditions, 1974.
Fourie, Elsje. 'A Future for the Theory of Multiple Modernities: Insights from the New Modernization Theory.' *Social Science Information* 51, 1 (2012): 52–69.

Frank, Roslyn M. 'Sociocultural Situatedness: An Introduction.' In *Sociocultural Situatedness*, edited by Roslyn M. Frank, René Dirven, Tom Ziemke, and Enrique Bernárdez, 1–18. Berlin: De Gruyter, 2008.

Franklin, Allan. *Shifting Standards: Experiments in Particle Physics in the Twentieth Century*. Pittsburgh, PA: University of Pittsburgh Press, 2013.

Fransen, J., D. Uebelhart, G. Stucki, T. Langenegger, M. Seitz, and B. A. Michel. 'The ICIDH-2 as a Framework for the Assessment of Functioning and Disability in Rheumatoid Arthritis.' *Annals of the Rheumatic Diseases* 61, 3 (2002): 225–31.

Franssen, Maarten, Pieter E. Vermaas, Peter Kroes, and Anthonie W. M. Meijers, eds. *Philosophy of Technology after the Empirical Turn*. New York: Springer, 2016.

Freeden, Michael. *Ideology: A Very Short Introduction*. Oxford: Oxford University Press, 2003.

Freeden, Michael. *Ideologies and Political Theory: A Conceptual Approach*. Repr. edn. Oxford: Clarendon Press, 2008.

Fricker, Miranda. 'Rational Authority and Social Power: Towards a Truly Social Epistemology.' *Proceedings of the Aristotelian Society* 98, 2 (1988): 159–77.

Friedman, Michael. 'Explanation and Scientific Understanding.' *Journal of Philosophy* 71 (1974): 5–19.

Friedman, Michael. 'Kuhn and the Rationality of Science.' *Philosophy of Science* 69, 2 (2002): 171–90.

Fuentes, Agustín and Aku Visala, eds. *Verbs, Bones, and Brains: Interdisciplinary Perspectives on Human Nature*. Notre Dame, IN: University of Notre Dame Press, 2017.

Fugelsang, Jonathan A. and Kevin N. Dunbar. 'A Cognitive Neuroscience Framework for Understanding Causal Reasoning and the Law.' *Philosophical Transactions of the Royal Society* 359 (2004): 1749–54.

Fuller, Peter. *Theoria: Art and the Absence of Grace*. London: Chatto & Windus, 1988.

Fumerton, Richard. 'Why You Can't Trust a Philosopher.' In *Disagreement*, edited by Richard Feldman and Ted A. Warfield, 91–111. Oxford: Oxford University Press, 2010.

Gabriel, Andrew K. *Barth's Doctrine of Creation: Creation, Nature, Jesus, and the Trinity*. Eugene, OR: Cascade Books, 2014.

Gale, Richard M. and Alexander R. Pruss. 'A New Cosmological Argument.' *Religious Studies* 35, 4 (1999): 461–76.

Galison, Peter. 'History, Philosophy, and the Central Metaphor.' *Science in Context* 2, 1 (1988): 197–212.

Galison, Peter. *Image and Logic: A Material Culture of Microphysics*. Chicago, IL: University of Chicago Press, 1997.

Galison, Peter. 'Material Culture, Theoretical Culture, and Delocalization.' In *Science in the Twentieth Century*, edited by John Krige and Dominique Pestre, 669–82. Amsterdam: Harwood, 1997.

Galison, Peter. 'Scientific Cultures.' In *Interpreting Clifford Geertz: Cultural Investigation in the Social Sciences*, edited by Jeffrey C. Alexander, Philip Smith, and Matthew Norton, 121–9. New York: Palgrave Macmillan, 2011.

Galison, Peter. 'The Journalist, the Scientist, and Objectivity.' In *Objectivity in Science: New Perspectives from Science and Technology Studies*, edited by Flavia Padovani, Alan Richardson, and Jonathan Y. Tsou, 57–75. Heidelberg: Springer, 2015.

Galison, Peter. 'Meanings of Scientific Unity: The Law, the Orchestra, the Pyramid, Quilt, and Ring.' In *Pursuing the Unity of Science: Ideology and Scientific Practice from the Great War to the Cold War*, edited by Harmke Kamminga and Geert Somsen, 12–29. Burlington, VT: Ashgate Publishing, 2016.

Gallagher, Shaun. *How the Body Shapes the Mind*. Oxford: Clarendon Press, 2005.

Galston, William A. 'Two Concepts of Liberalism.' *Ethics* 105, 3 (1995): 516–34.

Garber, Daniel. *Descartes Embodied: Reading Cartesian Philosophy through Cartesian Science*. Cambridge: Cambridge University Press, 2001.

Gattei, Stefano. *Thomas Kuhn's "Linguistic Turn" and the Legacy of Logical Empiricism: Incommensurability, Rationality and the Search for Truth*. Burlington, VT: Ashgate, 2008.

Gattei, Stefano. *Karl Popper's Philosophy of Science: Rationality without Foundations*. New York: Routledge, 2009.

Gaukroger, Stephen. *The Emergence of a Scientific Culture: Science and the Shaping of Modernity 1210–1685*. Oxford: Oxford University Press, 2006.

Gaukroger, Stephen. 'The Early Modern Idea of Scientific Doctrine and Its Early Christian Origins.' *Journal of Medieval and Early Modern Studies* 44, 1 (2014): 95–112.

Gebharter, Alexander. 'Causal Exclusion and Causal Bayes Nets.' *Philosophy and Phenomenological Research* 95, 2 (2017): 353–75.

Geertz, Clifford. *The Interpretation of Cultures: Selected Essays*. London: Fontana, 1973.

Geertz, Clifford. 'Common Sense as a Cultural System.' *Antioch Review* 33, 1 (1983): 5–26.

Gellner, Ernest. *Reason and Culture: The Historic Role of Rationality and Rationalism*. Oxford: Basil Blackwell, 1992.

Gemes, Ken. 'Hypothetico-Deductivism: Incomplete but Not Hopeless.' *Erkenntnis*. 63, 1 (2005): 139–47.

Geraci, Robert M. *Virtually Sacred: Myth and Meaning in World of Warcraft and Second Life.* New York: Oxford University Press, 2014.
Gervais, Will M., Aiyana K. Willard, Ara Norenzayan, and Joseph Henrich. 'The Cultural Transmission of Faith: Why Innate Intuitions Are Necessary, but Insufficient, to Explain Religious Belief.' *Religion* 41, 3 (2011): 389–410.
Gewirth, Alan. 'The Rationality of Reasonableness.' *Synthese* 57, 2 (1983): 225–47.
Ghosh, Suresh Chandra. '"English in Taste, in Opinions, in Words and Intellect:" Indoctrinating the Indian through Textbook, Curriculum and Education.' In *The Imperial Curriculum: Racial Images and Education in British Colonial Experience*, edited by J. A. Mangan, 175–93. London: Routledge, 1993.
Giere, Ronald N. *Science without Laws.* Chicago, IL: University of Chicago Press, 1999.
Giere, Ronald N. *Scientific Perspectivism.* Chicago, IL: University of Chicago Press, 2006.
Gieryn, Thomas F. 'Boundary-Work and the Demarcation of Science from Non-Science: Strains and Interests in Professional Ideologies of Scientists.' *American Sociological Review* 48 (1983): 781–95.
Gieryn, Thomas F. *Cultural Boundaries of Science: Credibility on the Line.* Chicago, IL: University of Chicago Press, 1999.
Gigerenzer, Gerd. 'The Adaptive Toolbox.' In *Bounded Rationality: The Adaptive Toolbox*, edited by Gerd Gigerenzer and Reinhard Selten, 37–50. Cambridge, MA: MIT Press, 2002.
Gliboff, Sander. 'Paley's Design Argument as an Inference to the Best Explanation, or, Dawkins' Dilemma.' *Studies in History and Philosophy of Biological and Biomedical Sciences* 31 (2000): 579–97.
Glover, Jonathan. *Humanity: A Moral History of the Twentieth Century.* London: Pimlico, 2001.
Gödel, Rainer. '"Eine unendliche Menge dunkeler Vorstellungen." Zur Widerstandigkeit von Empfindungen und Vorurteilen in der deutschen Spätaufklärung.' *Deutsche Vierteljahrsschrift für Literaturwissenschaft und Geistesgeschichte* 76, 4 (2002): 542–76.
Goetschel, Willi. *Spinoza's Modernity: Mendelssohn, Lessing, and Heine.* Madison, WI: University of Wisconsin Press, 2004.
Goldman, Alvin I. *Epistemology and Cognition.* Cambridge, MA: Harvard University Press, 1986.
Goldman, Alvin I. 'Group Knowledge versus Group Rationality: Two Approaches to Social Epistemology.' *Episteme: A Journal of Social Epistemology* 1, 1 (2004): 11–22.

Gonzalez, Wenceslao J. *La predicción científica: Concepciones filosófico-metodológicas desde H. Reichenbach a N. Rescher*. Barcelona: Montesinos, 2010.

Gooday, Graeme. 'Placing or Replacing the Laboratory in the History of Science?' *Isis* 99, 4 (2008): 783–95.

Goody, Jack. 'From Explanation to Interpretation in Social Anthropology.' In *Explanations: Styles of Explanation in Science*, edited by John Cornwell, 197–211. Oxford: Oxford University Press, 2004.

Gore, Charles. *The Incarnation of the Son of God*. London: John Murray, 1922.

Gould, Stephen Jay. *The Structure of Evolutionary Theory*. Cambridge, MA: Belknap Press, 2002.

Gozdecka, Dorota Anna. *Rights, Religious Pluralism and the Recognition of Difference*. New York: Routledge, 2016.

Grayling, A. C. *The God Argument*. London: Bloomsbury, 2013.

Gregory, Brad S. 'No Room for God? History, Science, Metaphysics, and the Study of Religion.' *History and Theory* 47, 4 (2008): 495–519.

Griese, Anneliese and Hans Jörg Sandkühler, eds. *Karl Marx—Zwischen Philosophie und Naturwissenschaften*. Frankfurt am Main: Peter Lang, 1997.

Grimmel, Andreas. 'Wittgenstein and the Context of Rationality: Towards a Language-Practical Notion of Rational Reason and Action.' *Journal of Language and Politics* 14, 5 (2015): 712–28.

Grosholz, Emily R. 'Descartes' Unification of Algebra and Geometry.' In *Descartes: Philosophy, Mathematics and Physics*, edited by Stephen Gaukroger, 156–68. Totowa, NJ: Barnes & Noble, 1980.

Grünbaum, Adolf. 'Is Falsifiability the Touchstone of Scientific Rationality? Karl Popper Versus Inductivism.' In *Essays in Memory of Imre Lakatos*, edited by R. S. Cohen, P. K. Feyerabend, and M. W. Wartofsky, 213–52. Dordrecht: Reidel, 1976.

Grünwald, Peter D. and Paul M. B. Vitányi. 'Kolmogorov Complexity and Information Theory.' *Journal of Logic, Language, and Information* 12, 4 (2003): 497–529.

Guarino, Thomas. 'Contemporary Theology and Scientific Rationality.' *Studies in Religion* 22 (1993): 311–22.

Guattari, Félix. 'La transversalité.' In *Psychanalyse et transversalité*, 72–85. Paris: La Découverte, 2003.

Gueth, Werner and Hartmut Kliemt. 'Perfect or Bounded Rationality? Some Facts, Speculations and Proposals.' *Analyse und Kritik—Zeitschrift für Sozialtheorie* 26 (2004): 364–81.

Gusdorf, Georges. *La conscience révolutionnaire: Les idéologues*. Paris: Payot, 1978.

Haack, Susan, *Defending Science—Within Reason: Between Scientism and Cynicism*. Amherst, MA: Prometheus, 2003.

Haas, Peter M. *Epistemic Communities, Constructivism, and International Environmental Politics*. London: Routledge, 2016.

Haidt, Jonathan. 'The Emotional Dog and Its Rational Tail: A Social Intuitionist Approach to Moral Judgment.' *Psychological Review* 108, 4 (2001): 814–34.

Hall, Catherine. 'Making Colonial Subjects: Education in the Age of Empire.' *History of Education* 37, 6 (2008): 773–87.

Hambrick, David Z. and Alexander P. Burgoyne. 'The Difference between Rationality and Intelligence.' *New York Times*, 16 September 2016.

Hammond, Kenneth R. *Beyond Rationality: The Search for Wisdom in a Troubled Time*. Oxford: Oxford University Press, 2007.

Hamou, Philippe. *La mutation du visible: Essai sur la portée épistémologique des instruments d'optique au XVIIe siècle*. Lille: Presses universitaires du Septentrion, 1999.

Hampshire, Stuart. *Innocence and Experience*. Cambridge, MA: Harvard University Press, 1989.

Hampshire, Stuart. *Justice Is Conflict*. Princeton, NJ: Princeton University Press, 2000.

Hannam, James. *God's Philosophers: How the Medieval World Laid the Foundations of Modern Science*. London: Icon, 2010.

Hanson, N. R. 'Is there a Logic of Scientific Discovery?' *Australasian Journal of Philosophy* 38 (1961): 91–106.

Hardwig, John. 'Epistemic Dependence.' *Journal of Philosophy* 82, 7 (1985): 335–49.

Harker, David. 'Accommodation and Prediction: The Case of the Persistent Head.' *British Journal for Philosophy of Science* 57 (2006): 309–21.

Harman, Gilbert. 'The Inference to the Best Explanation.' *Philosophical Review* 74 (1965): 88–95.

Harré, Rom and Carl Jensen, eds. *Beyond Rationality: Contemporary Issues*. Newcastle: Cambridge Scholars, 2011.

Harrison, Edward. 'The Redshift-Distance and Velocity-Distance Laws.' *Astrophysical Journal* 403, 1 (1993): 28–31.

Harrison, Peter. 'Physico-Theology and the Mixed Sciences: The Role of Theology in Early Modern Natural Philosophy.' In *The Science of Nature in the Seventeenth Century*, edited by Peter Anstey and John Schuster, 165–83. Dordrecht: Springer, 2005.

Harrison, Peter. '"Science" and "Religion:" Constructing the Boundaries.' *Journal of Religion* 86, 1 (2006): 81–106.

Harrison, Peter. 'Introduction.' In *The Cambridge Companion to Science and Religion*, edited by Peter Harrison, 1–18. Cambridge: Cambridge University Press, 2010.

Harrison, Peter. *The Territories of Science and Religion*. Chicago, IL: University of Chicago Press, 2015.

Harrison, Victoria. 'The Pragmatics of Defining Religion in a Multi-Cultural World.' *International Journal for Philosophy of Religion* 59 (2006): 133–52.

Haselton, Martie, Gregory A. Bryant, Andreas Wilke, David Frederick, Andrew Galperin, Willem E. Frankenhuis, and Tyler Moore. 'Adaptive Rationality: An Evolutionary Perspective on Cognitive Bias.' *Social Cognition* 27, 5 (2009): 733–63.

Hasker, William. 'Alston on the Rationality of Doxastic Practices.' *Faith and Philosophy* 27, 2 (2010): 205–11.

Haugeland, John. *Having Thought: Essays in the Metaphysics of Mind*. Cambridge, MA: Harvard University Press, 1998.

Hayes, Julie Candler. *Reading the French Enlightenment: System and Subversion*. Cambridge: Cambridge University Press, 1999.

Heelan, Patrick A. 'Nietzsche's Perspectivalism: A Hermeneutic Philosophy of Science.' In *Nietzsche, Epistemology, and Philosophy of Science: Nietzsche and the Sciences*, edited by Babette E. Babich, 203–20. Dordrecht: Springer, 1999.

Heikes, Deborah K. *Rationality and Feminist Philosophy*. London: Continuum, 2011.

Heikes, Deborah K. *Rationality, Representation, and Race*. London: Palgrave Macmillan, 2016.

Heisenberg, Werner. 'Naturwissenschaftliche und religiöse Wahrheit.' *Physikalische Blätter* 29, 8 (1973): 339–49.

Heisenberg, Werner. *Die Ordnung der Wirklichkeit*. Munich: Piper Verlag, 1989.

Heisenberg, Werner. 'Die Kopenhagener Deutung der Quantentheorie.' In his *Physik und Philosophie*, 67–85. Stuttgart: Hirzel, 2007.

Heller, Erich. *The Disinherited Mind*. New York: Meridian Books, 1959.

Hempel, Carl G. *Aspects of Scientific Explanation*. New York: Free Press, 1965.

Hempel, Carl G. and Paul Oppenheim. 'Studies in the Logic of Explanation.' *Philosophy of Science* 15, 2 (1948): 135–75.

Hempelmann, Heinzpeter. '"Keine ewigen Wahrheiten als unaufhörliche zeitliche": Hamanns Kontroverse mit Kant über Sprache und Vernunft.' *Theologische Beiträge* 18 (1987): 5–33.

Hendrix, John S. *Platonic Architectonics: Platonic Philosophies & the Visual Arts*. New York: Peter Lang, 2004.

Henry, John. 'National Styles in Science: A Factor in the Scientific Revolution?' In *Geography and Revolution*, edited by David N. Livingstone and Charles W. J. Withers, 43–74. Chicago, IL: University of Chicago Press, 2005.

Henry, John. 'Testimony and Empiricism: John Sergeant, John Locke, and the Social History of Truth.' In *Conflicting Values of Inquiry: Ideologies of Epistemology in Early Modern Europe*, edited by Tamás Demeter, Kathryn Murphy, and Claus Zittel, 95–124. Leiden: Brill, 2014.

Herdt, Jennifer A. 'Alasdair Macintyre's "Rationality of Traditions" and Tradition-Transcendental Standards of Justification.' *Journal of Religion* 78 (1998): 524–46.

Hesse, Hermann. '*Mit dem Erstaunen fängt es an:' Herkunft und Heimat; Natur und Kunst*. Frankfurt: Suhrkamp Verlag, 1986.

Hesse, Mary. 'In Defence of Objectivity.' *Proceedings of the British Academy* 58 (1972): 167–86.

Hesse, Mary. 'Models of Theory Change.' In *Logic, Methodology and Philosophy of Science*, edited by Patrick Suppes, Leon Henkin, and Athanese Joja, 379–91. Amsterdam: North Holland, 1973.

Hillman, Donald J. 'The Measurement of Simplicity.' *Philosophy of Science* 29, 3 (1962): 225–52.

Hinchman, Lewis P. and Sandra K. Hinchman, eds. *Memory, Identity, Community: The Idea of Narrative in the Human Sciences*. Albany, NY: State University of New York Press, 1997.

Hitchcock, Christopher and Elliott Sober. 'Prediction versus Accommodation and the Risk of Overfitting.' *British Journal for Philosophy of Science* 55 (2004): 1–34.

Hitchens, Christopher. *God Is Not Great: How Religion Poisons Everything*. New York: Twelve, 2007.

Hobbes, Thomas. *English Works*. 11 vols. London: Bohn, 1839.

Hon, Giora. 'Kepler's Revolutionary Astronomy: Theological Unity as a Comprehensive View of the World.' In *Conflicting Values of Inquiry: Ideologies of Epistemology in Early Modern Europe*, edited by Tamás Demeter, Kathryn Murphy, and Claus Zittel, 155–75. Leiden: Brill, 2014.

Hood, Ralph W., Peter C. Hill, and W. Paul Williamson. *The Psychology of Religious Fundamentalism*. New York: Guilford Press, 2005.

Hookway, Christopher. *Truth, Rationality, and Pragmatism: Themes from Peirce*. Oxford: Clarendon Press, 2000.

Hookway, Christopher. 'Interrogatives and Uncontrollable Abductions.' *Semiotica* 153 (2005): 101–15.

Höppner, Joachim. 'Karl Marx – Begründer des Wissenschaftlichen Kommunismus.' *Deutsche Zeitschrift für Philosophie* 31, 4 (1983): 389–403.

Hornbostel, Stefan. *Wissenschaftsindikatoren: Bewertungen in der Wissenschaft.* Opladen: Westdeutscher Verlag, 1997.
Horst, Steven W. *Beyond Reduction: Philosophy of Mind and Post-Reductionist Philosophy of Science.* Oxford: Oxford University Press, 2007.
Horst, Steven W. *Cognitive Pluralism.* Cambridge, MA: MIT Press, 2016.
Hountondji, Paulin J., ed. *La rationalité, une ou plurielle?* Dakar: Codesria, 2007.
Hoyningen-Huene, Paul. 'Context of Discovery and Context of Justification.' *Studies in History and Philosophy of Science* 18 (1987): 501–15.
Huggett, Nick. 'Local Philosophies of Science.' *Philosophy of Science* 67 (2000): S128–37.
Hull, David L. *Science and Selection: Essays on Biological Evolution and the Philosophy of Science.* Cambridge: Cambridge University Press, 2001.
Hume, David. *Letters*, edited by J. Y. T. Greig. 2 vols. Oxford: Clarendon Press, 1932.
Hume, David. *Essays, Moral, Political, and Literary*, ed. Eugene F. Miller. rev. edn: Indianapolis, IN: Liberty, 1987.
Hunt, Lynn and Margaret Jacob. 'Enlightenment Studies.' In *Encyclopedia of the Enlightenment*, edited by Alan Charles Kors, vol. 1, 418–30. 4 vols. Oxford: Oxford University Press, 2003.
Hurtado, Larry W. *One God, One Lord: Early Christian Devotion and Ancient Jewish Monotheism.* 3rd edn. London: Bloomsbury, 2015.
Hurtado, Larry W. *Ancient Jewish Monotheism and Early Christian Jesus-Devotion: The Context and Character of Christological Faith.* Waco, TX: Baylor University Press, 2017.
Iannaccone, Laurence R., Rodney Stark, and Roger Finke. 'Rationality and the "Religious Mind".' *Economic Inquiry* 36, 3 (1998): 373–89.
Illari, Phyllis. 'Mechanistic Explanation: Integrating the Ontic and Epistemic.' *Erkenntnis* 78 (2013): 237–55.
Illiffe, Rob. 'Newton, God, and the Mathematics of the Two Books.' In *Mathematicians and Their Gods: Interactions between Mathematics and Religious Beliefs*, edited by Snezana Lawrence and Mark McCartney, 121–44. Oxford: Oxford University Press, 2015.
Illiffe, Rob. *Priest of Nature: The Religious Worlds of Isaac Newton.* Oxford: Oxford University Press, 2017.
Immel, Oliver. 'Vom vernünftigen Ich: Überlegungen zur identitätsstiftenden Rolle der Rationalität im Anschluss an Jean-Paul Sartre.' In *Vernunft der Aufklärung—Aufklärung der Vernunft*, edited by Renate Reschke, Oliver Immel, Andreas Hütig, and Konstantin Broese, 365–80. Berlin: Akademie Verlag, 2006.

Immerwahr, John. 'Hume's Revised Racism.' *Journal of the History of Ideas* 53, 3 (1992): 481-6.

Inazu, John D. *Confident Pluralism: Surviving and Thriving through Deep Difference.* Chicago, IL: University of Chicago Press, 2016.

Inge, William R. *Faith and Its Psychology.* New York: Charles Scribner's Sons, 1910.

Israel, Jonathan I. *Enlightenment Contested: Philosophy, Modernity, and the Emancipation of Man, 1670-1752.* Oxford: Oxford University Press, 2008.

Ivanova, Milena. 'Poincaré's Aesthetics of Science.' *Synthese* 194 (2017): 2581-94.

Jammer, Max. *Einstein and Religion: Physics and Theology.* Princeton, NJ: Princeton University Press, 1999.

Janz, Paul D. *God, the Mind's Desire: Reference, Reason and Christian Thinking.* Cambridge: Cambridge University Press, 2004.

Jarvie, I. C. *Rationality and Relativism: In Search of a Philosophy and History of Anthropology.* London: Routledge, 2016.

Johnson, Maxwell E. *Praying and Believing in Early Christianity: The Interplay between Christian Worship and Doctrine.* Collegeville, MN: Liturgical Press, 2013.

Jones, James W. *Can Science Explain Religion? The Cognitive Science Debate.* New York: Oxford University Press, 2016.

Jones, Karen. 'Rationality and Gender.' In *The Oxford Handbook of Rationality*, edited by Alfred R. Mele and Piers Rawling, 301-19. Oxford: Oxford University Press, 2004.

Jong, Jonathan. 'On (Not) Defining (Non)Religion.' *Science, Religion and Culture* 2, 3 (2015): 15-24.

Jong, Jonathan, Christopher Kavanagh, and Aku Visala. 'Born Idolaters: The Limits of the Philosophical Implications of the Cognitive Science of Religion.' *Neue Zeitschrift für systematische Theologie und Religionsphilosophie* 57, 2 (2015): 244-66.

Joravsky, David. *Soviet Marxism and Natural Science, 1917-1932.* New York: Columbia University Press, 1961.

Jos, John T., Aaron C. Kay, and Hulda Thorisdottir. 'On the Social and Psychological Bases of Ideology.' In *Social and Psychological Bases of Ideology and System Justification*, edited by John T. Jos, Aaron C. Kay, and Hulda Thorisdottir, 3-23. Oxford: Oxford University Press, 2009.

Josephson-Storm, Jason A. *The Myth of Disenchantment: Magic, Modernity, and the Birth of the Human Sciences.* Chicago, IL: University of Chicago Press, 2017.

Jüngel, Eberhard. *Erfahrungen mit der Erfahrung: Unterwegs bemerkt.* Stuttgart: Radius Verlag, 2008.

Kagan, Jerome. *The Three Cultures: Natural Sciences, Social Sciences, and the Humanities in the 21st Century*. Cambridge: Cambridge University Press, 2009.

Kahneman, Daniel and Amos Tversky. 'Subjective Probability: A Judgment of Representativeness.' *Cognitive Psychology* 3 (1972): 430–54.

Kalanithi, Paul. *When Breath Becomes Air*. London: Vintage Books, 2017.

Kalberg, Stephen. 'Max Weber's Types of Rationality: Cornerstones for the Analysis of Rationalization Processes in History.' *American Journal of Sociology* 85, 5 (1980): 1145–79.

Kant, Immanuel. *Was ist Aufklärung?* Saillon: Jean Meslier Verlag, 2017.

Kaplan, Jonathan. 'Economic Rationality and Explaining Human Behavior: An Adaptationist Program?' *International Journal of Interdisciplinary Social Sciences* 7, 3 (2008): 79–94.

Karafyllis, Nicole Christine, ed. *Zugänge zur Rationalität der Zukunft*. Stuttgart/Weimar: Metzler, 2002.

Kärkkäinen, Veli-Matti.*Trinity and Religious Pluralism: The Doctrine of the Trinity in Christian Theology of Religions*. London: Routledge, 2017.

Kassler, Jamie C. *Newton's Sensorium: Anatomy of a Concept*. New York: Springer, 2018.

Kaufman, Daniel A. 'Knowledge, Wisdom, and the Philosopher.' *Philosophy* 81, 315 (2006): 129–51.

Kavanagh, Thomas M. 'Chance and Probability in the Enlightenment.' *French Forum* 15, 1 (1990): 5–24.

Keen, Sam. *Gabriel Marcel*. Richmond, VA: John Knox Press, 1967.

Keil, Frank C. 'Explanation and Understanding.' *Annual Review of Psychology* 57 (2006): 227–54.

Kekulé, August. 'Benzolfest Rede.' *Berichte der deutschen chemischen Gesellschaft zu Berlin* 23 (1890): 1302–11.

Kelly, Kevin. 'A New Solution to the Puzzle of Simplicity.' *Philosophy of Science* 74, 5 (2007): 561–73.

Keltner, Dacher and Jonathan Haidt. 'Approaching Awe, a Moral, Spiritual and Aesthetic Emotion.' *Cognition and Emotion* 17 (2003): 297–314.

Kenney, Catherine M. *The Remarkable Case of Dorothy L. Sayers*. Kent, OH: Kent State University Press, 1990.

Kenny, Anthony. *The Five Ways: St. Thomas Aquinas' Proofs of God's Existence*. London: Routledge & Kegan Paul, 2003.

Kenshur, Oscar. *Dilemmas of Enlightenment: Studies in the Rhetoric and Logic of Ideology*. Berkeley, CA: University of California Press, 1993.

Kepler, Johann. *Gesammelte Werke*, edited by Max Caspar. 22 vols. Munich: C. H. Beck, 1937–83.

Kern, Udo. *Dialektik der Vernunft bei Martin Luther*. Berlin–Münster: Lit, 2014.
Kidd, Ian James. 'Receptivity to Mystery: Cultivation, Loss, and Scientism.' *European Journal for Philosophy of Religion* 4, 3 (2012): 51–68.
Kidd, Ian James. 'Doing Science an Injustice: Midgley on Scientism.' In *Science and the Self: Animals, Evolution, and Ethics: Essays in Honour of Mary Midgley*, edited by Ian James Kidd and Liz McKinnell, 151–67. New York: Routledge, 2016.
Kidd, Ian James. 'Why Did Feyerabend Defend Astrology? Integrity, Virtue, and the Authority of Science.' *Social Epistemology* 30, 4 (2016): 464–82.
Kidd, Ian James. 'Reawakening to Wonder: Wittgenstein, Feyerabend, and Scientism.' In *Wittgenstein and Scientism*, edited by Ian James Kidd and Jonathan Beale, 101–15. London: Routledge, 2017.
Kidd, Ian James. 'Epistemic Vices in Public Debate: The Case of "New Atheism".' In *New Atheism: Critical Perspectives and Contemporary Debates*, edited by Christopher Cotter, Philip Quadrio, and Jonathan Tuckett, 51–68. Dordrecht: Springer, 2017.
Kieckhefer, Richard. 'The Specific Rationality of Medieval Magic.' *American Historical Review* 99 (1994): 813–36.
Kimmel, Michael. 'Properties of Cultural Embodiment: Lessons from the Anthropology of the Body.' In *Sociocultural Situatedness*, edited by Roslyn M. Frank, René Dirven, Tom Ziemke, and Enrique Bernárdez, 77–108. Berlin: De Gruyter, 2008.
Kincaid, Harold, John Dupré, and Alison Wylie, eds. *Value-Free Science? Ideals and Illusions*. Oxford: Oxford University Press, 2007.
Kingsley, Peter. *Ancient Philosophy, Mystery, and Magic: Empedocles and Pythagorean Tradition*. Oxford: Clarendon Press, 1997.
Kirklik, Alex and Peter Storkerson. 'Naturalizing Peirce's Semiotics.' In *Model-Based Reasoning in Science and Technology Abduction, Logic, and Computational Discovery*, edited by Lorenzo Magnani, Walter A. Carnielli, and Claudio Pizzi, 31–50. Berlin: Springer, 2010.
Kirkpatrick, Lee A. 'Religion Is Not an Adaptation.' In *Psychology, Religion, and Spirituality. Where God and Science Meet*, edited by Patrick McNamara, 173–93. Westport, CT: Praeger, 2006.
Kitcher, Philip. *Abusing Science: The Case against Creationism*. Cambridge, MA: MIT Press, 1982.
Kitcher, Philip. 'Explanatory Unification and the Causal Structure of the World.' In *Scientific Explanation*, edited by P. Kitcher and W. Salmon, 410–505. Minneapolis, MN: University of Minnesota Press, 1989.
Kitcher, Philip. 'On the Explanatory Role of Correspondence Truth.' *Philosophy and Phenomenological Research* 64 (2002): 346–64.

Kitcher, Philip. 'The Importance of Dewey for Philosophy (and for Much Else Besides).' In *Preludes to Pragmatism: Toward a Reconstruction of Philosophy*, 1–21. Oxford: Oxford University Press, 2012.

Kivy, Peter. 'Science and Aesthetic Appreciation.' *Midwest Studies in Philosophy* 16, 1 (1991): 180–95.

Klauk, Tobias. 'Is There Such a Thing as Narrative Explanation?' *Journal of Literary Theory* 10, 1 (2016): 110–38.

Klein, Heinz K. and Michael D. Myers. 'A Set of Principles for Conducting and Evaluating Interpretive Field Studies in Information Systems.' *MIS Quarterly* 23, 1 (1999): 67–93.

Klein, Julie Thompson. 'Discourses of Transdisciplinarity: Looking Back to the Future.' *Futures* 63 (2014): 68–74.

Knop, Julia. *Ecclesia Orans: Liturgie als Herausforderung für die Dogmatik*. Freiburg: Herder, 2012.

Knorr-Cetina, Karin. *The Manufacture of Knowledge: An Essay on the Constructivist and Contextual Nature of Science*. Oxford: Pergamon Press, 1981.

Knorr-Cetina, Karin. *Epistemic Cultures: How the Sciences Make Knowledge*. Cambridge, MA: Harvard University Press, 1999.

Kołakowski, Leszek and Stuart Hampshire, eds. *The Socialist Idea: A Reappraisal*. London: Weidenfeld & Nicolson, 1974.

Kolodny, Niko. 'Why Be Rational?' *Mind* 114 (2005): 509–63.

Kornblith, Hilary. *Knowledge and Its Place in Nature*. Oxford: Clarendon Press, 2002.

Kosso, Peter. 'The Omniscienter: Beauty and Scientific Understanding.' *International Studies in the Philosophy of Science* 16, 1 (2002): 39–48.

Kragh, Helge. *Conceptions of Cosmos from Myths to the Accelerating Universe: A History of Cosmology*. Oxford: Oxford University Press, 2007.

Kretzmann, Norman. *The Metaphysics of Creation: Aquinas's Natural Theology in Summa Contra Gentiles II*. Oxford: Clarendon Press, 1999.

Kriegel, Uriah. 'The New Mysterianism and the Thesis of Cognitive Closure.' *Acta Analytica* 18 (2003): 177–91.

Kroupa, Pavel. 'The Dark Matter Crisis: Falsification of the Current Standard Model of Cosmology.' *Publications of the Astronomical Society of Australia* 29 (2012): 395–433.

Kruglanski, Arie W. and Donna M. Webster. 'Motivated Closing of the Mind: "Seizing" and "Freezing".' *Psychological Review* 103, 2 (1996): 263–83.

Kuhn, Thomas S. 'Objectivity, Value Judgment, and Theory Choice.' In his *The Essential Tension: Selected Studies in the Scientific Tradition and Change*, 320–39. Chicago, IL: University of Chicago Press, 1977.

Kuhn, Thomas S. 'Logic of Discovery or Psychology of Research?' In his *The Essential Tension: Selected Studies in the Scientific Tradition and Change*, 266-92. Chicago, IL: University of Chicago Press, 1977.

Kuhn, Thomas S. 'Rationality and Theory Choice.' *Journal of Philosophy* 80 (1983): 563-70.

Kuhn, Thomas S. *The Road since Structure: Philosophical Essays, 1970-1993*. Chicago, IL: University of Chicago Press, 2000.

Kuipers, Theo A. F. 'Beauty, a Road to the Truth.' *Synthese* 131 (2002): 291-328.

Kullmann, Wolfgang. *Aristoteles und die moderne Wissenschaft*. Stuttgart: Steiner Verlag, 1998.

Kumar, Victor. 'To Walk Alongside: Myth, Magic, and Mind in the Golden Bough.' *Hau: Journal of Ethnographic Theory* 6, 2 (2016): 233-54.

Kurtz, Paul. 'Re-enchantment: A New Enlightenment.' *Free Inquiry Magazine* 24, 3 (April-May 2004).

Kvanvig, Jonathan L. 'Affective Theism and People of Faith.' *Midwest Studies in Philosophy* 37 (2013): 109-28.

Labbé, Yves. 'La souffrance: problème ou mystère.' *Nouvelle Revue Théologique* 116, 4 (1994): 513-29.

Labron, Tim. *Science and Religion in Wittgenstein's Fly-Bottle*. London: Bloomsbury, 2017.

Ladyman, James and Don Ross. *Every Thing Must Go: Metaphysics Naturalized*. Oxford: Oxford University Press, 2007.

Landsman, N. P. *Mathematical Topics between Classical and Quantum Mechanics*. New York: Springer, 1998.

Lang, T. J. *Mystery and the Making of a Christian Historical Consciousness: From Paul to the Second Century*. Berlin: De Gruyter, 2015.

Lange, Marc. 'The Apparent Superiority of Prediction to Accommodation as a Side Effect.' *British Journal for Philosophy of Science* 52 (2001): 575-88.

Largent, Mark A. 'Darwin's Analogy between Artificial and Natural Selection in the *Origin of Species*.' In *The Cambridge Companion to the 'Origin of Species'*, edited by Michael Ruse and Robert J. Richards, 12-29. Cambridge: Cambridge University Press, 2009.

Larsen, Eva L. and Christiane Stock. 'Capturing Contrasted Realities: Integrating Multiple Perspectives of Danish Community Life in Health Promotion.' *Health Promotion International* 26, 1 (2011): 14-22.

Latour, Bruno. 'From the World of Science to the World of Research?' *Science* 280, 5361 (1998): 208-9.

Latour, Bruno and Steven Woolgar. *Laboratory Life: The Construction of Scientific Facts*. 2nd edn. Princeton. NJ: Princeton University Press, 1986.

Laudan, Larry. *Science and Values: An Essay on the Aims of Science and Their Role in Scientific Debate.* Berkeley, CA: University of California Press, 1984.
Laudan, Larry and Jarrett Leplin. 'Empirical Equivalence and Underdetermination.' *Journal of Philosophy* 88 (1991): 449–72.
Lawrence, Snezana and Mark McCartney, eds. *Mathematicians and Their Gods: Interactions between Mathematics and Religious Beliefs.* Oxford: Oxford University Press, 2015.
Lawson, Anton E. 'What Is the Role of Induction and Deduction In Reasoning and Scientific Inquiry?' *Journal of Research in Science Teaching* 42, 6 (2005): 716–40.
Lawson, Hilary. *Closure: A Story of Everything.* London: Routledge, 2001.
Leavy, Patricia. *Essentials of Transdisciplinary Research: Using Problem-Centered Methodologies.* Walnut Creek, CA: Left Coast Press, 2011.
Lee, Raymond L. M. 'In Search of Second Modernity: Reinterpreting Reflexive Modernization in the Context of Multiple Modernities.' *Social Science Information* 47, 1 (2008): 55–69.
Leech, David and Aku Visala. 'The Cognitive Science of Religion: Implications for Theism?' *Zygon* 46, 1 (2011): 47–64.
Lefort, Claude. *The Political Forms of Modern Society: Bureaucracy, Democracy, Totalitarianism.* Cambridge, MA: MIT Press, 1986.
Lefrançois, Guy R. *Theories of Human Learning.* 3rd edn. Pacific Grove, CA: Brooks/Cole Publishers, 1995.
Leftow, Brian. 'The Ontological Argument.' In *The Oxford Handbook of Philosophy of Religion*, edited by William J. Wainwright, 80–115. Oxford: Oxford University Press, 2005.
Legare, Cristine, E. Margaret Evans, Karl S. Rosengren, and Paul L. Harris. 'The Coexistence of Natural and Supernatural Explanations across Cultures and Development.' *Child Development* 83, 3 (2012): 779–93.
Lehmkühler, Karsten. *Kultus und Theologie: Dogmatik und Exegese in der Religionsgeschichtlichen Schule.* Göttingen: Vandenhoeck & Ruprecht, 1996.
Leite, Adam J. 'On Justifying and Being Justified', *Noûs* 14 (2004): 219–53.
Leite, Adam J. 2008. 'Believing One's Reasons Are Good', *Synthese*, 161, 3 (2008): 419–41.
Lemeni, Adrian. 'The Rationality of the World and Human Reason as Expressed in the Theology of Father Dumitru Stăniloae: Points of Connection in the Dialogue between Theology and Science.' *International Journal of Orthodox Theology* 3, 4 (2012): 89–101.
Lenk, Hans. 'Typen und Systematik der Rationalität.' In *Kritik der wissenschaftlichen Rationalität*, edited by Hans Lenk, 11–27. Freiburg im Breisgau: Alber Verlag, 1986.

Leplin, Jarrett. 'Methodological Realism and Scientific Rationality.' *Philosophy of Science* 53 (1986): 31–51.
Levering, Matthew. *Scripture and Metaphysics: Aquinas and the Renewal of Trinitarian Theology*. Oxford: Blackwell, 2004.
Levi, Isaac. 'On the Seriousness of Mistakes.' *Philosophy of Science* 29 (1962): 47–65.
Levin, Yakir, Arnon Cahen, and Izhak Aharon. 'Naturalized Rationality, Evolutionary Psychology and Economic Theory.' *Journal of Cognition and Neuroethics* 1, 1 (2013): 39–72.
Levitin, Dmitri. 'Reconsidering John Sergeant's Attacks on Locke's *Essay*.' *Intellectual History Review* 20, 4 (2010): 457–77.
Lévy, Jacques, ed. *A Cartographic Turn: Mapping and the Spatial Challenge in Social Sciences*. Lausanne: EPFL Press, 2015.
Lewis, C. S. *An Experiment in Criticism*. Cambridge: Cambridge University Press, 1961.
Lewis, C. S. *Essay Collection*. London: HarperCollins, 2001.
Lewis, C. S. *Mere Christianity*. London: HarperCollins, 2002.
Li, Yan, Neal M. Ashkanasy, and David Ahlstrom. 'The Rationality of Emotions: A Hybrid Process Model of Decision-Making under Uncertainty.' *Asia Pacific Journal of Management* 31, 1 (2014): 293–308.
Lieu, Judith. *Christian Identity in the Jewish and Graeco-Roman World*. Oxford: Oxford University Press, 2006.
Lifschitz, Avi. *Language and Enlightenment: The Berlin Debates of the Eighteenth Century*. Oxford: Oxford University Press, 2012.
Lim, Paul Chang-Ha. *Mystery Unveiled: The Crisis of the Trinity in Early Modern England*. New York: Oxford University Press, 2012.
Lindbeck, George. *The Nature of Doctrine*. Philadelphia, PA: Westminster, 1984.
Lipton, Peter. *Inference to the Best Explanation*. 2nd edn. London: Routledge, 2004.
Lipton, Peter. 'What Good Is an Explanation?' In *Explanations: Styles of Explanation in Science*, edited by John Cornwell, 1–21. Oxford: Oxford University Press, 2004.
List, Christian. 'Group Knowledge and Group Rationality.' *Episteme: A Journal of Social Epistemology* 2, 1 (2005): 25–38.
List, Christian and Philip Pettit. 'Aggregating Sets of Judgments: An Impossibility Result.' *Synthese* 140 (2002): 207–35.
Livingstone, David N. 'Science, Text, and Space: Thoughts on the Geography of Reading.' *Transactions of the Institute of British Geographers* 35 (2005): 391–401.

Lloyd, Elisabeth Anne. 'The Nature of Darwin's Support for the Theory of Natural Selection.' In her *Science, Politics, and Evolution*, 1–19. Cambridge: Cambridge University Press, 2008.

Lloyd, G. E. R. *Magic, Reason and Experience: Studies in the Origin and Development of Greek Science*. Cambridge: Cambridge University Press, 1979.

Locke, John. *An Essay Concerning Human Understanding*, edited by P. H. Nidditch. Oxford: Oxford University Press, 1975.

Loke, Andrew Ter Ern. *God and Ultimate Origins: A Novel Cosmological Argument*. New York: Palgrave Macmillan, 2017.

Long, D. Stephen. *Speaking of God: Theology, Language, and Truth*. Grand Rapids, MI: Eerdmans, 2009.

López-Huertas, María. 'Reflections on Multidimensional Knowledge: Its Influence on the Foundation of Knowledge Organization.' *Knowledge Organization* 40, 6 (2013): 400–7.

Louth, Andrew. *Discerning the Mystery: An Essay on the Nature of Theology*. Oxford: Clarendon Press, 1983.

Louth, Andrew. *Origins of the Christian Mystical Tradition: From Plato to Denys*. Oxford: Oxford University Press, 2007.

Lowe, E. J. *The Four-Category Ontology: A Metaphysical Foundation for Natural Science*. Oxford: Oxford University Press, 2006.

Luhrmann, Tanya M., Howard Nusbaum, and Ronald Thisted. 'The Absorption Hypothesis: Learning to Hear God in Evangelical Christianity.' *American Anthropologist* 112, 1 (2010): 66–78.

Lycan, William G. *Judgement and Justification*. Cambridge: Cambridge University Press, 1988.

Lykke, Nina. 'Women's/Gender/Feminist Studies – A Post-Disciplinary Discipline?' In *The Making of European Women's Studies*, edited by Rosi Braidotti, Edyta Just, and Marlise Mensink, 91–101. Utrecht: Athena, 2004.

Lynham, Susan A. 'Theory Building in the Human Resource Development Profession.' *Human Resource Development Quarterly* 11, 2 (2000): 157–78.

Maasen, Sabine, Martin Lengwiler, and Michael Guggenheim. 'Practices of Transdisciplinary Research: Close(r) Encounters of Science and Society.' *Science and Public Policy* 33, 6 (2006): 393–6.

Machamer, Peter K. 'Activities and Causation: The Metaphysics and Epistemology of Mechanisms.' *International Studies in the Philosophy of Science* 18, 1 (2004): 27–39.

MacIntyre, Alasdair C. *After Virtue: A Study in Moral Theory*. 2nd edn. Notre Dame, IN: University of Notre Dame Press, 1984.

MacIntyre, Alasdair C. *Whose Justice? Which Rationality?* London: Duckworth, 1988.

MacIntyre, Alasdair C. *Dependent Rational Animals: Why Human Beings Need the Virtues*. Chicago, IL: Open Court, 1999.
MacKisack, Matthew, Susan Aldworth, Fiona Macpherson, John Onians, Crawford Winlove, and Adam Zeman. 'On Picturing a Candle: The Prehistory of Imagery Science.' *Frontiers in Psychology* 7, 515 (2016): 1–16.
MacMillan, Ken. 'Benign and Benevolent Conquest? The Ideology of Elizabethan Expansion Revisited.' *Early American Studies: An Interdisciplinary Journal* 9 (2011): 32–72.
Magnani, Lorenzo. *Abduction, Reason, and Science: Processes of Discovery and Explanation*. New York: Plenum Publishers, 2001.
Maher, Patrick. 'Prediction, Accommodation, and the Logic of Discovery.' *Philosophy of Science Association* 1 (1988): 273–85.
Maij, David L. R., Hein T. van Schie, and Michiel van Elk. 'The Boundary Conditions of the Hypersensitive Agency Detection Device: An Empirical Investigation of Agency Detection in Threatening Situations.' *Religion, Brain & Behavior* (2017): 1–29. doi: 10.1080/2153599X.2017.1362662.
Malcolm, Norman. *Wittgenstein: A Religious Point of View?* London: Routledge, 1993.
Mandelbrote, Scott. 'Eighteenth-Century Reactions to Newton's Anti-Trinitarianism.' In *Newton and Newtonianism: New Studies*, edited by J. E. Force and S. Hutton, 93–112. Dordrecht: Kluwer, 2004.
Mandelbrote, Scott. 'The Uses of Natural Theology in Seventeenth-Century England.' *Science in Context* 20 (2007): 451–80.
Marcel, Gabriel. *Being and Having*. London: Dacre Press, 1949.
Marcel, Gabriel. *The Philosophy of Existentialism*. New York: Citadel, 1995.
Marchand, Trevor H. J., ed. *Making Knowledge: Explorations of the Indissoluble Relation between Mind, Body and Environment*. Oxford: Wiley-Blackwell, 2010.
Margolis, Howard. *Patterns, Thinking, and Cognition: A Theory of Judgment*. Chicago, IL: University of Chicago Press, 1987.
Marion, Jean-Luc. 'The General Rule of Truth in the Third Meditation.' In *On the Ego and on God: Further Cartesian Questions*, 42–62. New York: Fordham University Press, 2007.
Markley, Robert. 'Objectivity as Ideology: Boyle, Newton, and the Languages of Science.' *Genre* 16 (1983): 355–72.
Márkus, György. 'Concepts of Ideology in Marx.' *Canadian Journal of Political and Social Theory* 15, 1 (1983): 87–106.
Marsden, John. 'Richard Tawney: Moral Theology and the Social Order.' *Political Theology* 7, 2 (2006): 181–99.
Marshall, Paul. 'Towards a Complex Critical Realism.' In *Metatheory for the Twenty-First Century: Critical Realism and Integral Theory in Dialogue*,

edited by Roy Bhaskar, Sean Esbjörn-Hargens, Nicholas Hedlund, and Mervyn Hartwig, 140–82. London: Routledge, 2016.
Martin, C. F. J. *Thomas Aquinas: God and Explanations*. Edinburgh: Edinburgh University Press, 1997.
Martín-Velasco, María José and María José García Blanco, eds. *Greek Philosophy and Mystery Cults*. Newcastle upon Tyne: Cambridge Scholars Publishing, 2016.
Maxwell, Nicholas. 'The Rationality of Scientific Discovery (I and II).' *Philosophy of Science* 41 2; 3 (1974): 123–53; 247–95.
Mazzotta, Giuseppe. *The New Map of the World: The Poetic Philosophy of Giambattista Vico*. Princeton, NJ: Princeton University Press, 1999.
McAllister, James W. 'Truth and Beauty in Scientific Reason.' *Synthese* 78 (1989): 25–51.
McAllister, James W. 'The Simplicity of Theories: Its Degree and Form.' *Journal for General Philosophy of Science* 22, 1 (1991): 1–14.
McCauley, Robert N. *Why Religion Is Natural and Science Is Not*. New York: Oxford University Press, 2011.
McGrath, Alister E. *Iustitia Dei: A History of the Christian Doctrine of Justification*. 3rd edn. Cambridge: Cambridge University Press, 2005.
McGrath, Alister E. 'Theologie als Mathesis Universalis? Heinrich Scholz, Karl Barth, und der wissenschaftliche Status der christlichen Theologie.' *Theologische Zeitschrift* 62 (2007): 44–57.
McGrath, Alister E. 'Gli ateismi di successo: Il nuovo Scientismo.' *Concilium: Rivista internazionale di teologia* 46, 4 (2010): 17–29.
McGrath, Alister E. *Darwinism and the Divine: Evolutionary Thought and Natural Theology*. Oxford: Wiley-Blackwell, 2011.
McGrath, Alister E. 'An Enhanced Vision of Rationality: C. S. Lewis on the Reasonableness of Christian Faith.' *Theology* 116, 6 (2013): 410–7.
McGrath, Alister E. 'Arrows of Joy: Lewis's Argument from Desire.' In his *The Intellectual World of C. S. Lewis*, 105–28. Oxford: Wiley-Blackwell, 2013.
McGrath, Alister E. 'A Gleam of Divine Truth: The Concept of Myth in Lewis's Thought.' In his *The Intellectual World of C. S. Lewis*, 55–82. Oxford: Wiley-Blackwell, 2013.
McGrath, Alister E. *C. S. Lewis—A Life. Eccentric Genius, Reluctant Prophet*. London: Hodder & Stoughton, 2013.
McGrath, Alister E. 'Hesitations about Special Divine Action: Reflections on Some Scientific, Cultural and Theological Concerns.' *European Journal for Philosophy of Religion* 7, 4 (2015): 3–22.
McGrath, Alister E. *Dawkins' God: From the Selfish Gene to the God Delusion*. 2nd edn. Oxford: Wiley-Blackwell, 2015.

McGrath, Alister E. 'The Rationality of Faith: How Does Christianity Make Sense of Things?' *Philosophia Christi* 18, 2 (2016): 395–408.
McGrath, Alister E. *Enriching our Vision of Reality: Theology and the Natural Sciences in Dialogue*. London: SPCK, 2016.
McGrath, Alister E. *Re-Imagining Nature: The Promise of a Christian Natural Theology*. Oxford: Wiley-Blackwell, 2016.
McGrath, Alister E. 'Chance and Providence in the Thought of William Paley.' In *Abraham's Dice: Chance and Providence in the Monotheistic Traditions*, edited by Karl Giberson, 240–59. Oxford: Oxford University Press, 2016.
McGrath, Alister E. 'Natürliche Theologie: Ein Plädoyer für eine neue Definition und Bedeutungserweiterung.' *Neue Zeitschrift für Systematische Theologie und Religionsphilosophie* 59, 3 (2017): 1–14.
McKaughan, Daniel J. 'From Ugly Duckling to Swan: C. S. Peirce, Abduction, and the Pursuit of Scientific Theories.' *Transactions of the Charles S. Peirce Society* 44, 3 (2008): 446–68.
McKenzie, Gérman. *Interpreting Charles Taylor's Social Theory on Religion and Secularization*. New York: Springer, 2016.
McManus, I. C. 'Symmetry and Asymmetry in Aesthetics and the Arts.' *European Review* 13, 2 (2005): 157–80.
McMullin, Ernan. 'Values in Science.' In *Proceedings of the 1982 Biennial Meeting of the Philosophy of Science Association*, edited by P. D. Asquith and T. Nickles, 3–28. East Lansing, MI: Philosophy of Science Association, 1983.
McMyler, Benjamin. *Testimony, Trust, and Authority*. Oxford: Oxford University Press, 2011.
Merritt, David. 'Cosmology and Convention.' *Studies in History and Philosophy of Modern Physics* 57 (2017): 41–52.
Mews, Constant J. 'The World as Text: The Bible and the Book of Nature in Twelfth-Century Theology.' In *Scripture and Pluralism: Reading the Bible in the Religiously Plural Worlds of the Middle Ages and Renaissance*, edited by Thomas J. Heffernan and Thomas E. Burman, 95–122. Leiden: Brill, 2005.
Middleton, J. Richard. *The Liberating Image: The Imago Dei in Genesis 1*. Grand Rapids, MI: Brazos Press, 2005.
Midgley, Mary. *Wisdom, Information, and Wonder: What Is Knowledge For?* London: Routledge, 1995.
Midgley, Mary. *Evolution as a Religion: Strange Hopes and Stranger Fears*. 2nd edn. London: Routledge, 2002.
Midgley, Mary. *The Myths We Live By*. London: Routledge, 2004.
Midgley, Mary. *Are You an Illusion?* Durham: Acumen, 2014.

Mignolo, Walter. *Desobediencia epistémica: Retórica de la modernidad, lógica de la colonialidad y gramática de la descolonialidad.* Buenos Aires: Ediciones del Signo, 2010.

Miller, Fred D. Jr. 'The Rule of Reason in Plato's *Laws*.' In *Reason, Religion, and Natural Law: From Plato to Spinoza*, edited by Jonathan A. Jacobs, 31–62. Oxford: Oxford University Press, 2012.

Miller, Jerome A. *In the Throe of Wonder: Intimations of the Sacred in a Post-Modern World.* Albany, NY: State University of New York Press, 1992.

Miller, Richard W. *Fact and Method: Explanation, Confirmation and Reality in the Natural and the Social Sciences.* Princeton, NJ: Princeton University Press, 1987.

Millican, Peter J. R. 'Hume's Sceptical Doubts Concerning Induction.' In *Reading Hume on Human Understanding*, edited by P. J. R. Millican, 107–73. Oxford: Clarendon Press, 2002.

Millon-Delsol, Chantal. 'La dénaturation de la vérité ou le fondement des idéologies.' *Laval théologique et philosophique* 44, 3 (1988): 339–43.

Minnameier, Gerhard. 'Peirce-Suit of Truth: Why Inference to the Best Explanation and Abduction Ought Not to Be Confused.' *Erkenntnis* 60, 1 (2004): 75–105.

Mitchell, Basil. *The Justification of Religious Belief.* London: Macmillan, 1973.

Moore, Andrew. *Realism and Christian Faith: God, Grammar, and Meaning.* Cambridge: Cambridge University Press, 2003.

Moore, A. W. 'Varieties of Sense-Making.' *Midwest Studies in Philosophy* 37 (2013): 1–10.

Morrison, Margaret. 'A Study in Theory Unification: The Case of Maxwell's Electromagnetic Theory.' *Studies in History and Philosophy of Science* 23 (1992): 103–45.

Morrison, Margaret. *Unifying Scientific Theories: Physical Concepts and Mathematical Structures.* Cambridge: Cambridge University Press, 2000.

Morrison, Margaret. 'One Phenomenon, Many Models: Inconsistency and Complementarity.' *Studies in History and Philosophy of Science Part A* 42, 2 (2011): 342–51.

Morrison, Margaret. *Reconstructing Reality: Models, Mathematics, and Simulations.* Oxford: Oxford University Press, 2015.

Mortimore, G. W. and J. B. Maund. 'Rationality in Belief.' In *Rationality and the Social Sciences*, edited by S. I. Benn and G. W. Mortimore, 11–33. London: Routledge, 1976.

Mouffe, Chantal. *The Return of the Political.* London: Verso, 2005.

Mulhall, Stephen. 'Wittgenstein on Faith, Rationality, and the Passions.' *Modern Theology* 27,2 (2011): 313–24.

Murdoch, Iris. *Metaphysics as a Guide to Morals.* London: Penguin, 1992.

Murphy, Kathryn and Anita Traninger, eds. *The Emergence of Impartiality*. Leiden: Brill, 2014.
Murphy, Nancey. *Theology in the Age of Scientific Reasoning*. Ithaca, NY: Cornell University Press, 1990.
Myers, Benjamin. 'The Stratification of Knowledge in the Thought of T. F. Torrance.' *Scottish Journal of Theology* 61 (2008): 1–15.
Nagel, Thomas. *The View from Nowhere*. New York: Oxford University Press, 1986.
Nagel, Thomas. 'Why Is There Anything?' In his *Secular Philosophy and the Religious Temperament: Essays, 2002–2008*, 27–32. Oxford: Oxford University Press, 2009.
Nersessian, Nancy J. 'How Do Scientists Think? Capturing the Dynamics of Conceptual Change in Science.' In *Cognitive Models of Science*, edited by Ronald N. Giere, 3–44. Minneapolis, MN: University of Minnesota Press, 1992.
Neuser, Wolfgang. *Natur und Begriff: Zur Theoriekonstitution und Begriffsgeschichte von Newton bis Hegel*. 2nd edn. Wiesbaden: Springer, 2017.
Newton-Smith, W. H. *The Rationality of Science*. London: Routledge, 1981.
Newton-Smith, W. H. 'Explanation.' In *A Companion to the Philosophy of Science*, edited by W. H. Newton-Smith, 127–33. Oxford: Blackwell, 2000.
Nicholas, Jeffery. *Reason, Tradition, and the Good: Macintyre's Tradition-Constituted Reason and Frankfurt School Critical Theory*. Notre Dame, IN: University of Notre Dame Press, 2012.
Nickerson, Raymond S. 'Teaching Reasoning.' In *The Nature of Reasoning*, edited by Robert J. Sternberg and Jacqueline P. Leighton, 410–42. Cambridge: Cambridge University Press, 2004.
Nickerson, Raymond S. *Aspects of Rationality: Reflections on What It Means to Be Rational and Whether We Are*. New York: Psychology Press, 2008.
Nickles, Thomas. 'Discovery.' In *Companion to the History of Modern Science*, edited by G. N. Cantor, J. R. R. Christie, M. J. S. Hodge, and R. C. Olby, 148–65. London: Routledge, 1990.
Niebuhr, H. Richard. *The Meaning of Revelation*. Louisville, KY: Westminster John Knox Press, 2006.
Nielsen, Marie Vejrup. *Sin and Selfish Genes: Christian and Biological Narratives*. Leuven: Peeters, 2010.
Nietzsche, Friedrich, *Sämtliche Werke: Kritische Studienausgabe*, edited by G. Colli and M. Montinari. 15 vols. Munich: DTV Verlagsgesellschaft, 1980–99.
Noble, Denis. 'A Theory of Biological Relativity: No Privileged Level of Causation.' *Interface Focus* 6, 2 (2012): 55–64.

Nozick, Robert. *The Nature of Rationality*. Princeton, NJ: Princeton University Press, 1993.
Numbers, Ronald L. 2010. 'Simplifying Complexity: Patterns in the History of Science and Religion.' In *Science and Religion: New Historical Perspectives*, edited by Thomas Dixon, G. N. Cantor, and Stephen Pumfrey, 263–82. Cambridge: Cambridge University Press, 2010.
Nye, Mary Jo. *Before Big Science: The Pursuit of Modern Chemistry and Physics, 1800–1940*. Cambridge, MA: Harvard University Press, 1999.
O'Brien, Timothy L. and Shiri Noy. 'Traditional, Modern, and Post-Secular Perspectives on Science and Religion in the United States.' *American Sociological Review* 80, 1 (2015): 92–115.
Ochs, Elinor and Lisa Capps. 'Narrating the Self.' *Annual Review of Anthropology* 25 (1996): 19–43.
Ochs, Peter. *Peirce, Pragmatism and the Logic of Scripture*. Cambridge: Cambridge University Press, 1998.
O'Connor, Timothy. *Theism and Ultimate Explanation: The Necessary Shape of Contingency*. Oxford: Wiley-Blackwell, 2012.
Oderberg, David S. 'Traversal of the Infinite, the "Big Bang" and the Kalām Cosmological Argument.' *Philosophia Christi* 4 (2002): 304–34.
O'Leary, Joseph Stephen. *Questioning Back: The Overcoming of Metaphysics in Christian Tradition*. Eugene, OR: Wipf & Stock, 2016.
O'Neill, Colman E. 'The Rule Theory of Doctrine and Propositional Truth.' *The Thomist* 49, 3 (1985): 417–82.
Ong, Walter J. 'The Shifting Sensorium.' In *The Varieties of Sensory Experience*, edited by David Howes, 47–60. Toronto: University of Toronto Press, 1991.
Operskalski, Joachim T. and Aron K. Barbey. 'Cognitive Neuroscience of Causal Reasoning.' In *The Oxford Handbook of Causal Reasoning*, edited by Michael Waldmann, 217–43. Oxford: Oxford University Press, 2017.
Ortega y Gasset, José. *Ideas y creencias*. Madrid: Alianza, 1997.
Osborne, Jonathan. 'Teaching Critical Thinking? New Directions in Science Education.' *School Science Review* 95, 352 (2014): 53–62.
Osborne, Jonathan, Shirley Simon, and Susan Collins. 'Attitudes towards Science: A Review of the Literature and Its Implications.' *International Journal of Science Education* 25, 9 (2003): 1049–79.
Osborne, Peter. 'Problematizing Disciplinarity, Transdisciplinary Problematics.' *Theory, Culture & Society* 32, 5–6 (2015): 3–35.
Ostrowick, John. 'Is Theism a Simple, and Hence, Probable, Explanation for the Universe?' *South African Journal of Philosophy* 31, 2 (2012): 354–68.

Otto, Rudolf. *Das Heilige: Über das Irrationale in der Idee des Göttlichen und sein Verhältnis zum Rationalen*. Gotha: Leopold Klotz, 1927.
Paavola, Sami. 'Abduction as a Logic of Discovery: The Importance of Strategies.' *Foundations of Science* 9 (2005): 267-83.
Paavola, Sami. 'Hansonian and Harmanian Abduction as Models of Discovery.' *International Studies in the Philosophy of Science* 20 (2006): 93-108.
Pagden, Anthony. *The Enlightenment and Why It Still Matters*. Oxford: Oxford University Press, 2013.
Pais, Abraham. *Niels Bohr's Times, in Physics, Philosophy and Polity*. Oxford: Clarendon Press, 1991.
Pannenberg, Wolfhart. *Systematic Theology*. 3 vols. Grand Rapids, MI: Eerdmans, 1991-8.
Pannenberg, Wolfhart. 'God as Spirit – and Natural Science.' *Zygon* 36, 4 (2001): 783-94.
Panofsky, Erwin. 'Die Perspektive als "Symbolische Form".' In his *Aufsätze zu Grundfragen der Kunstwissenschaft*, 99-167. Berlin: Volker Spiess, 1980.
Park, Crystal L. 'Religion as a Meaning-Making Framework in Coping with Life Stress.' *Journal of Social Issues* 61, 4 (2005): 707-29.
Peczenik, Aleksander. 'A Theory of Legal Doctrine.' *Ratio Juris* 14, 1 (2001): 75-105.
Peirce, Charles S. *Collected Papers*, edited by Charles Hartshorne and Paul Weiss. 8 vols. Cambridge, MA: Harvard University Press, 1960.
Pelikan, Jaroslav. *Christianity and Classical Culture: The Metamorphosis of Natural Theology in the Christian Encounter with Hellenism*. New Haven, CT: Yale University Press, 1993.
Pennock, Robert T. *Tower of Babel: The Evidence against the New Creationism*. Cambridge, MA: MIT Press, 1999.
Penrose, Roger. *The Road to Reality: A Complete Guide to the Laws of the Universe*. London: Jonathan Cape, 2004.
Penrose, Roger. *Fashion, Faith, and Fantasy in the New Physics of the Universe*. Princeton, NJ: Princeton University Press, 2017.
Peters, James R. *The Logic of the Heart: Augustine, Pascal, and the Rationality of Faith*. Grand Rapids, MI: Baker Academic, 2009.
Pettit, Philip. 'Groups with Minds of Their Own.' In *Socializing Metaphysics: The Nature of Social Reality*, edited by Frederick Schmitt, 167-93. Lanham, MD: Rowman and Littlefield, 2003.
Pettit, Philip. 'When to Defer to Majority Testimony – and When Not.' *Analysis* 66, 3 (2006): 179-87.
Pew Research Center and the AAAS. 'Public Praises Science; Scientists Fault Public, Media Scientific Achievements Less Prominent Than a Decade Ago.' Washington, DC: Pew Research Center, 2009.

Pham, Michel Tuan. 'Emotion and Rationality: A Critical Review and Interpretation of Empirical Evidence.' *Review of General Psychology* 11, 2 (2007): 155–78.

Phillips, Dewi Z. *Religion without Explanation*. Oxford: Blackwell, 1976.

Pickstock, Catherine. *After Writing: On the Liturgical Consummation of Philosophy*. Oxford: Blackwell Publishers, 1997.

Pietsch, Wolfgang. 'Defending Underdetermination or Why the Historical Perspective Makes a Difference.' In *EPSA Philosophy of Science: Amsterdam 2009*, edited by Henk W. de Regt, Stephan Hartmann, and Samir Okasha, 303–13. Dordrecht: Springer, 2012.

Piff, Paul K., Pia Dietze, Matthew Feinberg, Daniel M. Stancato, and Dacher Keltner. 'Awe, the Small Self, and Prosocial Behavior.' *Journal of Personality and Social Psychology* 108, 6 (2015): 883–99.

Pigliucci, Massimo. 'New Atheism and the Scientistic Turn in the Atheism Movement.' *Midwest Studies in Philosophy* 37, 1 (2013): 142–53.

Pigliucci, Massimo. 'Feyerabend and the Cranks: A Response to Shaw.' *Social Epistemology Review and Reply Collective* 6, 7 (2017): 1–6.

Pinker, Steven. *The Blank Slate: The Modern Denial of Human Nature*. New York: Viking, 2002.

Planck, Max. *Where Is Science Going?* New York: W. W. Norton, 1932.

Plantinga, Alvin. *The Nature of Necessity*. Oxford: Oxford University Press, 1974.

Plantinga, Alvin. *Warranted Christian Belief*. Oxford: Oxford University Press, 2000.

Plantinga, Alvin. *Where the Conflict Really Lies: Science, Religion, and Naturalism*. New York: Oxford University Press, 2011.

Plantinga, Alvin. *Knowledge and Christian Belief*. Grand Rapids, MI: Eerdmans, 2015.

Plutynski, Anya. 'Parsimony and the Fisher-Wright Debate.' *Biology and Philosophy* 20, 4 (2005): 697–713.

Plutynski, Anya. 'Explanatory Unification and the Early Synthesis.' *British Journal for Philosophy of Science* 56 (2005): 595–609.

Pojman, Louis. *Who Are We? Theories of Human Nature*. New York: Oxford University Press, 2006.

Polanyi, Michael. *Personal Knowledge: Towards a Post-Critical Philosophy*. London: Routledge, 1958.

Polkinghorne, John. *Science and Christian Belief*. London: SPCK, 1994.

Polkinghorne, John. *Scientists as Theologians: A Comparison of the Writings of Ian Barbour, Arthur Peacocke and John Polkinghorne*. London: SPCK, 1996.

Polkinghorne, John. 'Wolfhart Pannenberg's Engagement with the Natural Sciences.' *Zygon* 34, 1 (1999): 151–8.

Polkinghorne, John. 'Fields and Theology: A Response to Wolfhart Pannenberg.' *Zygon* 36, 4 (2001): 795–7.
Polkinghorne, John. 'Physics and Metaphysics in a Trinitarian Perspective.' *Theology and Science* 1 (2003): 33–49.
Polkinghorne, John. *Theology in the Context of Science*. London: SPCK, 2008.
Pope, Alexander. *Essay on Man*. Oxford: Oxford University Press, 1881.
Popper, Karl R. *Conjectures and Refutations: The Growth of Scientific Knowledge*. London: Routledge, 1963.
Popper, Karl R. 'Natural Selection and the Emergence of Mind.' *Dialectica* 32 (1978): 339–55.
Popper, Karl R. 'Truth without Authority.' In *Popper Selections*, edited by David Miller, 46–57. Princeton, NJ: Princeton University Press, 1985.
Popper, Karl R. *The Logic of Scientific Discovery*. New York: Routledge, 2002.
Portugal, Agnaldo Cuoco. 'Plantinga and the Bayesian Justification of Beliefs.' *Veritas* 57, 2 (2012): 15–25.
Potochnik, Angela. 'Levels of Explanation Reconceived.' *Philosophy of Science* 77, 1 (2010): 59–72.
Powell, Samuel M. *The Trinity in German Thought*. Cambridge: Cambridge University Press, 2008.
Prendinger, Helmut and Mitsuru Ishizuka. 'A Creative Abduction Approach to Scientific and Knowledge Discovery.' *Knowledge-Based Systems* 18 (2005): 321–6.
Preston, John. *Feyerabend: Philosophy, Science and Society*. Oxford: Wiley-Blackwell, 1997.
Prevost, Robert. *Probability and Theistic Explanation*. Oxford: Clarendon Press, 1990.
Prinz, Jesse J. *Beyond Human Nature: How Culture and Experience Shape the Human Mind*. New York: W. W. Norton, 2012.
Psillos, Stathis. *Scientific Realism: How Science Tracks Truth*. London: Routledge, 1999.
Putnam, Hilary. *Naturalism, Realism, and Normativity*. Cambridge, MA: Harvard University Press, 2016.
Pyenson, Lewis. *Empire of Reason: Exact Sciences in Indonesia, 1840–1940*. Leiden: Brill, 1989.
Quackenbush, Stephen L. *International Conflict: Logic and Evidence*. Los Angeles, CA: Sage, 2015.
Quante, Michael, David P. Schweikard, and Matthias Hoesch. *Marx-Handbuch: Leben—Werk—Wirkung* Stuttgart: Metzler Verlag, 2016.
Quash, Ben. *Found Theology: History, Imagination and the Holy Spirit*. London: Bloomsbury, 2013.

Quine, W. V. O. 'Two Dogmas of Empiricism.' In his *From a Logical Point of View*, 2nd edn, 20–46. Cambridge, MA: Harvard University Press, 1951.
Quine, W. V. O. 'On Simple Theories of a Complex World.' *Synthese* 15, 1 (1963): 103–6.
Radder, Hans. *The World Observed, the World Conceived*. Pittsburgh, PA: University of Pittsburgh Press, 2006.
Ramsey, Ian T. *Models and Mystery*. London: Oxford University Press, 1964.
Raphael, Melissa. *Rudolf Otto and the Concept of Holiness*. Oxford: Clarendon Press, 1997.
Rausch, Hannelore. *Theoria: Von ihrer sakralen zur philosophischen Bedeutung*. Munich: Fink, 1982.
Rawls, John. *Political Liberalism*. New York: Columbia University Press, 2005.
Raz, Joseph. *From Normativity to Responsibility*. Oxford: Oxford University Press, 2011.
Reeves, Josh. 'Problems for Postfoundationalists: Evaluating J. Wentzel Van Huyssteen's Interdisciplinary Theory of Rationality.' *Journal of Religion* 93, 2 (2013): 131–50.
Reich, K. Helmut. 'The Doctrine of the Trinity as a Model for Structuring the Relations Between Science and Theology.' *Zygon* 30 (1995): 383–405.
Reichenbach, Bruce R. 'Explanation and the Cosmological Argument.' In *Contemporary Debates in the Philosophy of Religion*, edited by Michael Peterson and Raymond van Arragon, 97–114. Oxford: Wiley-Blackwell, 2004.
Reichenbach, Hans. *The Rise of Scientific Philosophy*. Berkeley, CA: University of California Press, 1954.
Reisner, Andrew. 'Is There Reason to Be Theoretically Rational?' In *Reasons for Belief*, edited by Andrew Reisner and Asbjørn Steglich-Petersen, 34–53. Cambridge: Cambridge University Press, 2011.
Rennie, Bryan S. 'The View of the Invisible World: Ninian Smart's Analysis of the Dimensions of Religion and of Religious Experience.' *Bulletin of the Council of Societies for the Study of Religion* 28, 3 (1999): 63–9.
Rescher, Nicholas. *Rationalität: Eine philosophische Untersuchung über das Wesen und die Rechtfertigung von Vernunft*. Würzburg: Königshausen und Neumann, 1993.
Rescher, Nicholas. *Rationalität, Wissenschaft und Praxis*. Würzburg: Königshausen und Neumann, 2002.
Rescher, Nicholas. *Common Sense: A New Look at an Old Tradition*. Milwaukee, WI: Marquette University Press, 2005.
Rich, Patricia. 'Axiomatic and Ecological Rationality: Choosing Costs and Benefits.' *Erasmus Journal for Philosophy and Economics* 9, 2 (2016): 90–122.

Richmond, James. *Theology and Metaphysics.* London: SCM Press, 1970.
Riedweg, Christoph. *Mysterienterminologie bei Platon, Philon, und Klemens von Alexandrien.* Berlin: De Gruyter, 1987.
Rieger, Hans-Martin. 'Grenzen wissenschaftlicher Rationalität, Relativismus und Gottesglaube: Reflexionen zur zeitgenössischen Wissenschaftstheoretischen Diskussion.' *Theologische Beiträge* 33 (2002): 334–55.
Rivera, Nelson. *The Earth is Our Home: Mary Midgley's Critique and Reconstruction of Evolution and Its Meanings.* Exeter: Imprint Academic, 2010.
Roberts, Robert Campbell and W. Jay Wood. *Intellectual Virtues: An Essay in Regulative Epistemology.* Oxford: Clarendon Press, 2009.
Robinson, Andrew. *God and the World of Signs: Trinity, Evolution, and the Metaphysical Semiotics of C. S. Peirce.* Leiden: Brill, 2010.
Robinson, Daniel N. and Richard N. Williams, eds. *Scientism: The New Orthodoxy.* London: Bloomsbury, 2014.
Robinson, Dominic. *Understanding the 'Imago Dei': The Thought of Barth, von Balthasar and Moltmann.* London: Routledge, 2016.
Robinson, Marilynne. *What Are We Doing Here?* New York: Farrar, Straus and Giroux, 2018.
Rocha, Maristela do Nascimento and Ivã Gurgel. 'Descriptive Understandings of the Nature of Science: Examining the Consensual and Family Resemblance Approaches.' *Interchange* 48, 4 (2017): 403–29.
Rodrigues, Cassiano Terra. 'The Method of Scientific Discovery in Peirce's Philosophy: Deduction, Induction, and Abduction.' *Logica Universalis* 5 (2011): 127–64.
Roets, Arne and Alain van Hiel. 'Allport's Prejudiced Personality Today: Need for Closure as the Motivated Cognitive Basis of Prejudice.' *Current Directions in Psychological Science* 20, 6 (2011): 349–54.
Rogers, G. A. J. 'Stillingfleet, Locke and the Trinity.' In *Judaeo-Christian Intellectual Culture in the Seventeenth Century: A Celebration of the Library of Narcissus Marsh (1638-1713)*, edited by Allison P. Coudert, Sarah Hutton, Richard H. Popkin, and Gordon M. Weiner, 207–24. Dordrecht: Springer, 1999.
Roque, Alicia Juarrero. 'Language Competence and Tradition-Constituted Rationality.' *Philosophy and Phenomenological Research* 51 (1991): 611–17.
Rose, Steven. 'The Biology of the Future and the Future of Biology'. In *Explanations: Styles of Explanation in Science*, edited by John Cornwell, 125–42. Oxford: Oxford University Press, 2004.
Rosen, Gideon. 'Nominalism, Naturalism, Philosophical Relativism.' *Philosophical Perspectives* 15 (2001): 69–91.

Rosenberg, Alexander. *The Atheist's Guide to Reality: Enjoying Life without Illusions*. New York: W. W. Norton, 2011.
Rott, Hans and Verena Wagner. 'Das Ende vom Problem des methodischen Anfangs: Descartes' antiskeptisches Argument.' In *Homo Sapiens und Homo Faber: Epistemische und technische Rationalität in Antike und Gegenwart*, edited by Gereon Wolters and Jürgen Mittelstraß, 133–45. Berlin: De Gruyter, 2005.
Rouse, Joseph. *Knowledge and Power: Towards a Political Philosophy of Science*. Ithaca, NY: Cornell University Press, 1987.
Rouse, Joseph. *Engaging Science: How to Understand Its Practices*. Ithaca, NY: Cornell University Press, 1996.
Rouse, Joseph. *How Scientific Practices Matter: Reclaiming Philosophical Naturalism*. Chicago, IL: University of Chicago Press, 2002.
Rouse, Joseph. *Articulating the World: Conceptual Understanding and the Scientific Image*. Chicago, IL: The University of Chicago Press, 2015.
Rowan, Michael. 'Stove on the Rationality of Induction and the Uniformity Thesis.' *British Journal for the Philosophy of Science* 44, 3 (1993): 561–6.
Rueger, Alexander. 'Aesthetic Appreciation of Experiments: The Case of 18th-Century Mimetic Experiments.' *International Studies in the Philosophy of Science* 16, 1 (2002): 49–59.
Rueger, Alexander. 'Perspectival Models and Theory Unification.' *British Journal for the Philosophy of Science* 56 (2005): 579–94.
Rupke, Nicolaas A. 'A Geography of Enlightenment: The Critical Reception of Alexander von Humboldt's Mexico Work.' In *Geography and Enlightenment*, edited by Charles W. J. Withers and David N. Livingstone, 319–39. Chicago, IL: University of Chicago Press, 1999.
Rupert, Mark, ed. *Ideologies of Globalization: Contending Visions of a New World Order*. New York: Routledge, 2000.
Rupert, Robert. 'Minding One's Cognitive Systems: When Does a Group of Minds Constitute a Single Cognitive Unit?' *Episteme* 1, 3 (2005): 177–88.
Ruskin, John. *Works*. Edited by E. T. Cook and A. Wedderburn. 39 vols. London: Allen, 1903–12.
Russell, Bertrand. *Common Sense and Nuclear Warfare*. George Allen and Unwin, London, 1959.
Russell, Bertrand. *Bertrand Russell speaks his Mind*. London: Barker, 1960.
Russell, Bertrand. *History of Western Philosophy*. 2nd edn. London: George Allen & Unwin, 1961.
Russell, Bertrand. *Essays in Skepticism*. New York: Philosophical Library, 1963.
Russell, Bertrand. *Unpopular Essays*. New York: Routledge, 1996.

Russell, Bertrand and Frederick Copleston. 'Debate on the Existence of God.' In *The Existence of God*, edited by John Hick, 167–90. New York: Macmillan, 1964.

Russell, Colin A. 'The Conflict Metaphor and Its Social Origins.' *Science and Christian Faith* 1 (1989): 3–26.

Russell, Jeffrey Burton. *Inventing the Flat Earth: Columbus and Modern Historians*. New York: Praeger, 1991.

Rutzou, Timothy. 'Integral Theory and the Search for the Holy Grail: On the Possibility of a Metatheory.' *Journal of Critical Realism* 13, 1 (2014): 77–83.

Ryan, Sharon. 'Wisdom, Knowledge and Rationality.' *Acta Analytica* 27, 2 (2012): 99–112.

Saler, Benson. *Conceptualizing Religion: Immanent Anthropologists, Transcendent Natives and Unbounded Categories*. Leiden: Brill, 1993.

Saler, Benson. 'Theory and Criticism: The Cognitive Science of Religion.' *Method & Theory in the Study of Religion* 22, 4 (2010): 330–9.

Salmon, Wesley C. *Scientific Explanation and the Causal Structure of the World*. Princeton: Princeton University Press, 1984.

Salmon, Wesley C. *Four Decades of Scientific Explanation*. Minneapolis, MN: University of Minnesota Press, 1989.

Salmon, Wesley C. *Causality and Explanation*. Oxford: Oxford University Press, 1998.

Samuels, Richard and Stephen P. Stich. 'Rationality and Psychology.' In *The Oxford Handbook of Rationality*, edited by Alfred R. Mele and Piers Rawling, 279–300. Oxford: Oxford University Press, 2004.

Sanchez, Pascal. *La rationalité des croyances magiques*. Geneva: Droz, 2007.

Sánchez-Blanco, Francisco. *Michael Servets Kritik an der Trinitätslehre: Philosophische Implikationen und historische Auswirkungen*. Frankfurt am Main: Lang, 1977.

Sandel, Michael J. *Liberalism and the Limits of Justice*. 2nd edn. Cambridge, UK; New York: Cambridge University Press, 1998.

Sandler, Willibald. 'Christentum als große Erzählung: Anstöße für eine narrative Theologie.' In *Religion—Literatur—Künste: Ein Dialog*, edited by Peter Tschuggnall, 523–38. Anif: Müller-Speiser 2002.

Santos, Laurie R. and Michael L. Platt. 'Evolutionary Anthropological Insights into Neuroeconomics: What Non-Human Primates Can Tell Us About Human Decision-Making Strategies.' In *Neuroeconomics: Decision Making and the Brain*, edited by Paul W. Glimcher and Ernst Fehr, 109–22. Cambridge, MA: Academic Press, 2013.

Sayers, Dorothy L. *The Mind of the Maker*. London: Methuen, 1941.

Sayers, Dorothy L. *The Letters of Dorothy L. Sayers: Volume II, 1937 to 1943*, edited by Barbara Reynolds. New York: St Martin's Press, 1996.

Sayers, Dorothy L. *Child and Woman of Her Time: A Supplement to the Letters of Dorothy L. Sayers, Volume Five*, edited by Barbara Reynolds. Cambridge: Dorothy L. Sayers Society, 2002.

Schaefer, Donovan. 'Blessed, Precious Mistakes: Deconstruction, Evolution, and New Atheism in America.' *International Journal for Philosophy of Religion* 76 (2014): 75–94.

Schaffalitzky de Muckadell, Caroline. 'On Essentialism and Real Definitions of Religion.' *Journal of the American Academy of Religion* 82, 2 (2014): 495–520.

Schaffner, Kenneth F. *Discovery and Explanation in Biology and Medicine*. Chicago, IL: University of Chicago Press, 1993.

Schatzki, Theodore R., K. Knorr-Cetina, and Eike von Savigny, eds. *The Practice Turn in Contemporary Theory*. New York: Routledge, 2001.

Schickore, Jutta and Friedrich Steinle, eds. *Revisiting Discovery and Justification: Historical and Philosophical Perspectives on the Context Distinction*. New York: Springer, 2006.

Schiemann, Gregor. 'Welt im Wandel: Werner Heisenbergs Ansätze zu einer pluralistischen Philosophie.' In *Das bunte Gewand der Theorie: Vierzehn Begegnungen mit philosophierenden Forschern*, edited by Astrid Schwarz, 296–320. Freiburg: Alber, 2009.

Schilbrack, Kevin. 'Rationality, Relativism, and Religion: A Reinterpretation of Peter Winch.' *Sophia* 48, 4 (2009) 48: 399–412.

Schilbrack, Kevin. *Philosophy and the Study of Religions: A Manifesto*. Oxford: Wiley Blackwell, 2014.

Schliesser, Eric. 'Newton's Challenge to Philosophy.' *HOPOS: The Journal of the International Society for the History of Philosophy of Science* 1 (2011): 101–28.

Schlosshauer, Maximilian, Johannes Kofler, and Anton Zeilinger. 'A Snapshot of Foundational Attitudes toward Quantum Mechanics.' *Studies in the History and Philosophy of Modern Physics* 44 (2013): 220–30.

Schmidt, James. 'Inventing the Enlightenment: Anti-Jacobins, British Hegelians, and the *Oxford English Dictionary*.' *Journal of the History of Ideas* 64, 3 (2003): 421–43.

Schmidt, Volker H. 'Multiple Modernities or Varieties of Modernity?' *Current Sociology* 54, 1 (2006): 77–97.

Schmitt, Frederick. 'On the Road to Social Epistemic Interdependence.' *Social Epistemology* 2, 4 (1988): 297–307.

Schneider, Michael. *Zur theologischen Grundlegung des christlichen Gottesdienstes nach Joseph Ratzinger*. Cologne: Patristisches Zentrum Koinonia-Oriens, 2009.
Scholte, Bob. 'On the Ethnocentricity of Scientistic Logic.' *Dialectical Anthropology* 3, 2 (1978): 177–89.
Scholz, Oliver R. *Verstehen und Rationalität*. 3rd edn. Frankfurt am Main: Vittorio Klostermann, 2016.
Schrag, Calvin O. *The Resources of Rationality: A Response to the Postmodern Challenge*. Bloomington, IN: Indiana University Press, 1992.
Schumacher, Lydia. *Divine Illumination: The History and Future of Augustine's Theory of Knowledge*. Oxford: Wiley-Blackwell, 2011.
Schurz, Gerhard. 'Patterns of Abduction.' *Synthese*, 164, 2 (2008): 201–34.
Schüz, Peter. *Mysterium Tremendum: Zum Verhältnis von Angst und Religion nach Rudolf Otto*. Tübingen: Mohr Siebeck, 2016.
Schweder, Rebecca. 'A Defense of a Unificationist Theory of Explanation.' *Foundations of Science* 10 (2005): 421–35.
Seachris, Joshua. 'The Meaning of Life as Narrative: A New Proposal for Interpreting Philosophy's "Primary" Question.' *Philo* 12, 1 (2009): 5–23.
Searle, John. 'Rationality and Realism: What Is at Stake?' *Daedalus* 122 (1993): 55–83.
Segerstrale, Ullica. 'Wilson and the Unification of Science.' *Annals of the New York Academy of Sciences*, 1093 (2006): 46–73.
Seipel, Peter. 'Tradition-Constituted Inquiry and the Problem of Tradition-Inherence.' *The Thomist* 78, 3 (2014): 419–46.
Seipel, Peter. 'In Defense of the Rationality of Traditions.' *Canadian Journal of Philosophy* 45, 3 (2015): 257–77.
Sellars, Wilfrid. *Science, Perception and Reality*. New York: Humanities Press, 1962.
Șerban, Maria. 'What Can Polysemy Tell Us About Theories of Explanation?' *European Journal for Philosophy of Science* 7, 1 (2017): 45–56.
Shadish, William R., Thomas D. Cook, and Donald T. Campbell. *Experimental and Quasi-Experimental Designs for Generalized Causal Inference*. Boston, MA: Houghton Mifflin, 2001.
Shapere, Dudley. *Reason and the Search for Knowledge: Investigations in the Philosophy of Science*. Dordrecht: Springer, 1984.
Shapin, Steven and Simon Schaffer. *Leviathan and the Air-Pump: Hobbes, Boyle and the Experimental Life*. Princeton, NJ: Princeton University Press, 1985.
Shapiro, Lawrence A. *Embodied Cognition*. New York: Routledge, 2011.

Shea, Nicholas. 'Distinguishing Top-Down from Bottom-Up Effects.' In *Perception and Its Modalities*, edited by Dustin Stokes, Mohan Matthen, and Stephen Biggs, 73–91. Oxford: Oxford University Press, 2015.

Shea, William. 'Looking at the Moon as Another Earth. Terrestrial Analogies and Seventeenth-Century Telescopes.' In *Metaphors and Analogy in the Sciences*, edited by Fernand Hallyn, 83–104. Dordrecht: Kluwer Academic, 2000.

Shiota, Michelle N., Dacher Keltner, and Amanda Mossman. 'The Nature of Awe: Elicitors, Appraisals, and Effects on Self-Concept.' *Cognition and Emotion* 21, 5 (2007): 944–63.

Shults, F. LeRon. *The Postfoundationalist Task of Theology: Wolfhart Pannenberg and the New Theological Rationality*. Grand Rapids, MI: Eerdmans, 1999.

Shults, F. LeRon. *Reforming Theological Anthropology after the Philosophical Turn to Relationality*. Grand Rapids, MI: Eerdmans, 2003.

Shults, F. LeRon, ed. *The Evolution of Rationality: Interdisciplinary Essays in Honor of J. Wentzel Van Huyssteen*. Grand Rapids, MI: Eerdmans, 2006.

Sibley, W. M. 'The Rational Versus the Reasonable.' *Philosophical Review* 62 (1953): 554–60.

Sider, Robert D. 'Credo Quia Absurdum?' *Classical World* 73 (1978): 417–19.

Sider, Theodore. *Writing the Book of the World*. Oxford: Oxford University Press, 2012.

Siegel, Harvey. 'Truth, Problem Solving and the Rationality of Science.' *Studies in History and Philosophy of Science* 14 (1983): 89–112.

Simon, Herbert A. 'Rationality as a Process and as a Product of Thought.' *American Economic Review* 68 (1978): 1–16.

Simon, Herbert A. *Reason in Human Affairs*. Stanford, CA: Stanford University Press, 1983.

Sinhababu, Neil. 'The Humean Theory of Practical Irrationality.' *Journal of Ethics and Social Philosophy* 6, 1 (2011): 1–13.

Skinner, B. F. 'A Case History in the Scientific Method.' *American Psychologist* 11 (1956): 221–33.

Sloan, Phillip R. 'Darwin, Vital Matter, and the Transformism of Species.' *Journal of the History of Biology* 19 (1986): 369–445.

Sloman, Steven A. *Causal Models: How People Think About the World and Its Alternatives*. Oxford: Oxford University Press, 2005.

Smith, Christian. *Moral, Believing Animals: Human Personhood and Culture*. Oxford: Oxford University Press, 2009.

Smith, Christian. *Religion: What It Is, How It Works, and Why It Is Still Important*. Princeton, NJ: Princeton University Press, 2017.

Smith, Christian and Brandon Vaidyanathan. 'Multiple Modernities and Religion.' In *The Oxford Handbook of Religious Diversity*, edited by Chad Meister, 250-65. New York: Oxford University Press, 2010.

Smith, Robin. 'Logic.' In *The Cambridge Companion to Aristotle*, edited by Jonathan Barnes, 27-65. Cambridge: Cambridge University Press, 1999.

Snobelen, Stephen D. 'To Discourse of God: Isaac Newton's Heterodox Theology and His Natural Philosophy.' In *Science and Dissent in England, 1688-1945*, edited by Paul B. Wood, 39-65. Aldershot: Ashgate, 2004.

Snyder, Laura J. 'The Mill-Whewell Debate: Much Ado About Induction.' *Perspectives on Science* 5 (1997): 159-98.

Snyder, Laura J. *Reforming Philosophy: A Victorian Debate on Science and Society*. Chicago, IL: University of Chicago Press. 2006.

Sober, Elliot. 'What Is the Problem of Simplicity?' In *Simplicity, Inference and Modelling: Keeping It Sophisticatedly Simple*, edited by A. Zellner, H. A. Keuzenkamp, and M. McAleer, 13-31. Cambridge: Cambridge University Press, 2001.

Solomon, Julie R. *Objectivity in the Making: Francis Bacon and the Politics of Inquiry*. Baltimore, MD: Johns Hopkins University Press, 2002.

Solomon, Miriam. *Social Empiricism*. Cambridge, MA: MIT Press, 2007.

Sorabji, Richard. 'Rationality.' In *Rationality in Greek Thought*, edited by Michael Frede and Gisela Striker, 311-33. Oxford: Oxford University Press, 1996.

Sosa, Ernest. *Knowledge in Perspective: Selected Essays in Epistemology*. Cambridge: Cambridge University Press, 1991.

Sosa, Ernest and Michael Tooley. *Causation*. Oxford: Oxford University Press, 1993.

Southwold, Martin. 'Buddhism and the Definition of Religion.' *Man: New Series* 13, 3 (1978): 362-79.

Spaemann, Robert. 'Rationality and Faith in God.' *Communio* 32 (2005): 618-36.

Sparks, Adam. 'The Fulfilment Theology of Jean Daniélou, Karl Rahner and Jacques Dupuis.' *New Blackfriars* 89 (2008): 633-56.

Spranzi, Marta. 'Galileo and the Mountains of the Moon: Analogical Reasoning, Models and Metaphors in Scientific Discovery.' *Journal of Cognition and Culture* 4, 3-4 (2004): 451-83.

Sprenger, Jan. 'Hypothetico-Deductive Confirmation.' *Philosophy Compass* 6, 7 (2011): 497-508.

Stanovich, Keith E. *Rationality and the Reflective Mind*. New York: Oxford University Press, 2011.

Stanovich, Keith E. 'On the Distinction between Rationality and Intelligence: Implications for Understanding Individual Differences in Reasoning.' In *The Oxford Handbook of Thinking and Reasoning*, edited by Keith J. Holyoak and Robert G. Morrison, 433–55. Oxford: Oxford University Press, 2012.

Stanovich, Keith E. and Richard F. West. 'Individual Differences in Reasoning: Implications for the Rationality Debate?' *Behavioral and Brain Sciences* 23 (2000): 645–726.

Stanovich, Keith E. and Richard F. West. 'On the Relative Independence of Thinking Biases and Cognitive Ability.' *Journal of Personality and Social Psychology* 94, 4 (2008): 672–95.

Steger, Manfred B. *The Rise of the Global Imaginary: Political Ideologies from the French Revolution to the Global War on Terror*. Oxford: Oxford University Press, 2009.

Stenhouse, John. 'Darwin's Captain: F. W. Hutton and the Nineteenth-Century Darwinian Debates.' *Journal of the History of Biology* 23 (1990): 411–42.

Stenmark, Mikael. *Rationality in Science, Religion, and Everyday Life: A Critical Evaluation of Four Models of Rationality*. Notre Dame, IN: University of Notre Dame Press, 1995.

Stenmark, Mikael. *Scientism: Science, Ethics and Religion*. Aldershot: Ashgate, 2001.

Stepin, Vyacheslav S. 'Historical Types of Scientific Rationality.' *Russian Studies in Philosophy* 53, 2 (2015): 168–80.

Stewart, Alan. 'Bribery, Buggery, and the Fall of Lord Chancellor Bacon.' In *Rhetoric and Law in Early Modern Europe*, edited by Victoria Kahn and Lorna Hutson, 125–42. New Haven, CT: Yale University Press, 2001.

Stratton, S. Brian. *Coherence, Consonance, and Conversation: The Quest of Theology, Philosophy, and Natural Science for a Unified World-View*. Lanham, MD: University Press of America, 2000.

Strawson, Peter F. *Skepticism and Naturalism: Some Varieties*. New York: Columbia University Press, 1985.

Strevens, Michael. 'The Causal and Unification Approaches to Explanation Unified—Causally.' *Noûs* 38, 1 (2004): 154–76.

Stump, David J. 'Pierre Duhem's Virtue Epistemology.' *Studies in History and Philosophy of Science Part A* 38, 1 (2007): 149–59.

Sweetman, Brendan. *The Vision of Gabriel Marcel: Epistemology, Human Person, the Transcendent*. New York: Rodopi, 2008.

Swinburne, Richard. *Simplicity as Evidence for Truth*. Milwaukee, WI: Marquette University Press, 1997.

Swinburne, Richard, ed. *Bayes's Theorem*. Oxford: Oxford University Press, 2002.

Swinburne, Richard. *The Existence of God*. 2nd edn. Oxford: Clarendon Press, 2004.
Tambiah, Stanley. *Magic, Science and the Scope of Rationality*. Cambridge: Cambridge University Press, 1990.
Tanesini, Alessandra. 'Nietzsche's Theory of Truth.' *Australasian Journal of Philosophy* 73, 4 (1995): 548–59.
Tang, C. L. *Fundamentals of Quantum Mechanics*. Cambridge: Cambridge University Press, 2005.
Tanzella-Nitti, Giuseppe. 'La dimensione cristologica dell'intelligibilità del reale.' In *L'intelligibilità del reale: Natura, Uomo, Macchina*, edited by Sergio Rondinara, 213–25. Rome: Città Nuova, 1999.
Tanzella-Nitti, Giuseppe. 'Le rôle des sciences naturelles dans le travail du théologien.' *Revue des Questions Scientifiques* 170 (1999): 25–39.
Tanzella-Nitti, Giuseppe. 'Theologia physica: Razionalità scientifica e domanda su Dio.' *Hermeneutica: Annuario di filosofia e teologia* (2012): 37–54.
Tappolet, Christine. 'Emotions and the Intelligibility of Akratic Action.' In *Weakness of Will and Practical Irrationality*, edited by Sarah Stroud and Christine Tappolet, 97–120. Oxford: Clarendon Press, 2007.
Taylor, Charles. 'Modernity and Difference.' In *Without Guarantees: In Honour of Stuart Hall*, edited by Paul Gilroy, Lawrence Grossberg, and Angela McRobbie, 364–74. London: Verso, 2000.
Taylor, Charles. *Modern Social Imaginaries*. Durham, NC: Duke University Press, 2004.
Taylor, Charles. 'Geschlossene Weltstrukturen in der Moderne.' In *Wissen und Weisheit: Zwei Symposien zu Ehre von Josef Pieper*, edited by Hermann Fechtrup, Friedbert Schulze, and Thomas Sternberg, 137–69. Münster: LIT Verlag, 2005.
Taylor, Charles. *A Secular Age*. Cambridge, MA: Belknap Press, 2007.
Teller, Paul. 'Twilight of the Perfect Model Model.' *Erkenntnis* 55 (2001): 393–415.
ter Hark, Michel. 'Searching for the Searchlight Theory: From Karl Popper to Otto Selz.' *Journal of the History of Ideas* 64, 3 (2003): 465–97.
Thagard, Paul R. 'The Best Explanation: Criteria for Theory Choice.' *Journal of Philosophy* 75 (1978): 76–92.
Thagard, Paul R. *How Scientists Explain Disease*. Princeton, NJ: Princeton University Press, 1999.
Thagard, Paul R. 'Rationality and Science.' In *The Oxford Handbook of Rationality*, edited by Alfred R. Mele and Piers Rawling, 363–79. Oxford: Oxford University Press, 2004.

Thagard, Paul R. 'Why is Beauty a Road to Truth?' In *Cognitive Structures in Scientific Inquiry*, edited by R. Festa, A. Aliseda, and J. Pejnenburg, 365–70. Amsterdam: Rodopi, 2005.

Thagard, Paul R. 'Coherence, Truth, and the Development of Scientific Knowledge.' *Philosophy of Science* 74 (2007): 28–47.

Thiemann, Ronald F. *Constructing a Public Theology: The Church in a Pluralistic Culture*. Louisville, KY: Westminster John Knox Press, 1991.

Thomas, Keith. *Religion and the Decline of Magic: Studies in Popular Beliefs in Sixteenth and Seventeenth England*. London: Weidenfeld & Nicholson, 1971.

Thomas, Paul. *Marxism and Scientific Socialism: From Engels to Althusser*. London: Routledge, 2008.

Tilley, Terrence W. 'Ian Ramsey and Empirical Fit.' *Journal of the American Academy of Religion* 45, 3 (1977): G: 963–88.

Tobin, Theresa. 'Toward an Epistemology of Mysticism: Knowing God as Mystery.' *International Philosophical Quarterly* 50, 2 (2010): 221–41.

Todd, Cain S. 'Unmasking the Truth beneath the Beauty: Why the Supposed Aesthetic Judgements Made in Science May Not Be Aesthetic at All.' *International Studies in the Philosophy of Science* 22 (2008): 61–79.

Toplak, Maggie E., Richard F. West, and Keith E. Stanovich. 'Assessing the Development of Rationality.' In *The Developmental Psychology of Reasoning and Decision-Making*, edited by Henry Markovits, 7–35. New York: Psychology Press, 2014.

Torrance, Thomas F. *Theological Science*. London: Oxford University Press, 1969.

Torrance, Thomas F. *God and Rationality*. London: Oxford University Press, 1971.

Torrance, Thomas F. *Theology in Reconstruction*. Grand Rapids, MI: Eerdmans, 1996.

Toulmin, Stephen. *Human Understanding: The Collective Use and Evolution of Concepts*. Princeton, NJ: Princeton University Press, 1972.

Toulmin, Stephen. *Cosmopolis: The Hidden Agenda of Modernity*. New York: Free Press, 1990.

Toulmin, Stephen. 'Be Reasonable, Not Certain.' *Concepts and Transformation* 5, 2 (2000): 151–63.

Toulmin, Stephen. *The Uses of Argument*. Cambridge: Cambridge University Press, 2003.

Townley, Barbara. *Reason's Neglect: Rationality and Organizing*. Oxford: Oxford University Press, 2008.

Trenery, David. *Alasdair Macintyre, George Lindbeck, and the Nature of Tradition*. Eugene, OR: Pickwick Publications, 2014.

Trigg, Roger. *Rationality and Science: Can Science Explain Everything?* Oxford: Blackwell, 1993.
Trigg, Roger. *Rationality and Religion.* Oxford: Blackwell, 1998.
Trout, D. J. 'Scientific Explanation and the Sense of Understanding.' *Philosophy of Science* 69, 2 (2002): 212–33.
Tuomela, Raimo. *The Philosophy of Sociality: The Shared Point of View.* Oxford: Oxford University Press, 2007.
Turner, Frank Miller. 'The Victorian Conflict between Science and Religion: A Professional Dimension.' *Isis* 69 (1978): 356–76.
Üstün, T. B., S. Chatterji, J. Bickenbach, N. Kostanjsek, and M. Schneider. 'The International Classification of Functioning, Disability and Health: A New Tool for Understanding Disability and Health.' *Disability and Rehabilitation* 25 (2003): 565–71.
Valcárcel, Vicente Fernández. *Desengaños filosóficos.* 4 vols. Madrid: Don Blas Roman, 1787–97.
van Bavel, Tarsicius J. 'God in between Affirmation and Negation According to Augustine.' In *Augustin: Presbyter Factus Sum,* edited by Joseph T. Lienhard, Earl C. Muller, and Roland J. Teske, 73–97. New York: Peter Lang, 1993.
van den Torren, Benno. 'Distinguishing Doctrine and Theological Theory: A Tool for Exploring the Interface between Science and Faith.' *Science and Christian Belief* 28 (2016): 55–73.
van Fraassen, Bas. 'The Pragmatics of Explanation.' *American Philosophical Quarterly* 14 (1977): 143–50.
van Fraassen, Bas. *The Scientific Image.* Oxford: Oxford University Press, 1980.
van Fraassen, Bas. *The Empirical Stance.* New Haven, CT: Yale University Press, 2002.
van Fraassen, Bas. *Scientific Representation: Paradoxes of Perspective.* Oxford: Clarendon Press, 2010.
van Holten, Wilko. 'Does Religion Explain Anything? D. Z. Phillips and the "Wittgensteinian Objection" to Religious Explanation.' *Neue Zeitschrift für systematische Theologie und Religionsphilosophie* 44 (2002): 199–217.
van Holten, Wilko. 'Theism and Inference to the Best Explanation.' *Ars Disputandi* 2, 1 (2002): 262–81.
van Huyssteen, J. Wentzel. *Essays in Postfoundationalist Theology.* Grand Rapids, MI: Eerdmans, 1997.
van Huyssteen, J. Wentzel. *The Shaping of Rationality: Toward Interdisciplinarity in Theology and Science.* Grand Rapids, MI: Eerdmans, 1999.
van Huyssteen, J. Wentzel. *Alone in the World? Human Uniqueness in Science and Theology.* Grand Rapids, MI: Eerdmans, 2006.

van Waarden, Frans and Michaela Drahos. 'Courts and (Epistemic) Communities in the Convergence of Competition Policies.' *Journal of European Public Policy* 6 (2002): 913–4.
Vico, Giambattista. *La scienza nuova*. Milan: Biblioteca universale Rizzoli, 1977.
Vietta, Silvio. *Rationalität: Eine Weltgeschichte. Europäische Kulturgeschichte und Globalisierung*. Paderborn: Fink, 2012.
Visala, Aku. 'Explaining Religion at Different Levels: From Fundamentalism to Pluralism.' In *The Roots of Religion: Exploring the Cognitive Science of Religion*, edited by Roger Trigg and Justin L. Barrett, 55–74. London: Routledge, 2016.
Vogel, Jonathan. 'Epistemic Bootstrapping.' *Journal of Philosophy* 105, 9 (2008): 518–39.
von Helmholtz, Hermann. *Über die Erhaltung der Kraft: Eine physikalische Abhandlung*. Berlin: Reimer, 1847.
Vorster, Jakobus M. 'Perspectives on the Core Characteristics of Religious Fundamentalism Today.' *Journal for the Study of Religions and Ideologies* 21, 7 (2008): 44–65.
Votsis, Ioannis. 'Putting Realism in Perspective.' *Philosophica* 84 (2012): 85–122.
Waap, Thorsten. *Gottebenbildlichkeit und Identität: Zum Verhältnis von theologischer Anthropologie und Humanwissenschaft bei Karl Barth und Wolfhart Pannenberg*. Göttingen: Vandenhoeck & Ruprecht, 2008.
Wainwright, William J. 'Worldviews, Criteria and Epistemic Circularity.' In *Inter-Religious Models and Criteria*, edited by James Kellenberger, 87–105. London: Palgrave Macmillan, 1993.
Walker, Michael T. 'The Social Construction of Mental Illness and Its Implications for the Recovery Model.' *International Journal of Psychosocial Rehabilitation* 10, 1 (2006): 71–87.
Walsh, Dorothy. 'Occam's Razor: A Principle of Intellectual Elegance.' *American Philosophical Quarterly* 16, 3 (1979): 241–4.
Ward, Graham. *Unbelievable: Why We Believe and Why We Don't*. London: I.B. Tauris, 2014.
Watson, Francis. *Text, Church and World: Biblical Interpretation in Theological Perspective*. London: T&T Clark, 2004.
Watson, Nicholas. 'The Trinitarian Hermeneutic in Julian of Norwich's *Revelation of Divine Love*.' In *Julian of Norwich: A Book of Essays*, edited by Sandra J. McEntire, 61–90. New York: Garland Publishing, 1998.
Watts, Fraser. 'Are Science and Religion in Conflict?' *Zygon* 32 (1997): 125–38.
Watts, Fraser. 1998. 'Science and Theology as Complementary Perspectives.' In *Rethinking Theology and Science: Six Models for the Current Dialogue*,

edited by Niels Henrik Gregersen and J. Wentzel van Huyssteen, 157–79. Grand Rapids, MI: Eerdmans, 1998.

Webel, Charles. *The Politics of Rationality: Reason through Occidental History*. New York: Routledge, 2014.

Weinberg, Steven. *The First Three Minutes: A Modern View of the Origin of the Universe*. New York: Harper, 1993.

Weinberg, Steven. 'Can Science Explain Everything? Can Science Explain Anything?' In *Explanations: Styles of Explanation in Science*, edited by John Cornwell, 23–38. Oxford: Oxford University Press, 2004.

Weinstein, Philip. 'The View from Somewhere.' *Raritan* 32, 4 (2013): 85–101.

Weisberg, Michael. *Simulation and Similarity: Using Models to Understand the World*. New York: Oxford University Press, 2013.

Welsch, Wolfgang. 'Nietzsche über Vernunft – "Meine Wiederhergestellte Vernunft".' In *Rationalität und Prärationalität*, edited by Jan Beaufort and Peter Prechtl, 107–15. Würzburg: Königshausen & Neumann, 1998.

Welsch, Wolfgang. *Vernunft: Die zeitgenössische Vernunftkritik und das Konzept der Transversalen Vernunft*. 4th edn. Frankfurt am Main: Suhrkamp, 2007.

Welshon, Robert C. 'Perspectivist Ontology and *De Re* Knowledge.' In *Nietzsche's Philosophy of Science: Reflecting Science on the Ground of Art and Life*, edited by Babette E. Babich, 39–46. Albany, NY: State University of New York Press, 1994.

Weyl, Hermann. *Symmetry*. Princeton, NJ: Princeton University Press, 1952.

Whewell, William. *Philosophy of the Inductive Sciences*. 2 vols. London: Parker, 1847.

Whewell, William. *On the Philosophy of Discovery: Chapters Historical and Critical*. London: Parker, 1860.

White, Roger. 'The Epistemic Advantage of Prediction over Accommodation.' *Mind* 112 (2003): 653–83.

Whitehead, Alfred North. *The Concept of Nature*. Cambridge: Cambridge University Press, 1920.

Wigner, Eugene. 'The Unreasonable Effectiveness of Mathematics.' *Communications on Pure and Applied Mathematics* 13 (1960): 1–14.

Wilber, Ken. *Sex, Ecology, Spirituality: The Spirit of Evolution*. Boston, MA: Shambhala Publications, 1995.

Williams, R. J. P., Allan Chapman, and J. S. Rowlinson. *Chemistry at Oxford: A History from 1600 to 2005*. Cambridge: Royal Society of Chemistry, 2009.

Williams, Rowan. *Arius: Heresy and Tradition*. Grand Rapids, MI: Eerdmans, 2002.

Williamson, Timothy. *Knowledge and Its Limits*. Oxford: Oxford University Press, 2000.

Wilson, Edward O. *Sociobiology: The New Synthesis*. Cambridge, MA: Harvard University Press, 1975.
Wilson, Edward O. 'Resuming the Enlightenment Quest.' *Wilson Quarterly* 22, 1 (1998): 16–27.
Wilson, Edward O. *Consilience: The Unity of Knowledge*. New York: Vintage, 1999.
Winch, Peter. 'Understanding a Primitive Society.' *American Philosophical Quarterly* 1, 4 (1964): 307–24.
Winsberg, Eric, Bryce Huebner, and Rebecca Kukla. 'Accountability and Values in Radically Collaborative Research.' *Studies in History and Philosophy of Science Part A* 46, 1 (2014): 16–23.
Witham, Larry. *Marketplace of the Gods: How Economics Explains Religion*. Oxford: Oxford University Press, 2010.
Withers, Charles W. J. *Placing the Enlightenment: Thinking Geographically about the Age of Reason*. Chicago, IL: University of Chicago Press, 2007.
Wittgenstein, Ludwig. *On Certainty*. Oxford: Blackwell, 1974.
Wittgenstein, Ludwig. *Tractatus Logico-Philosophicus*. London: Routledge, 1992.
Wittgenstein, Ludwig. *Philosophical Investigations*. 4th edn. Oxford: Wiley-Blackwell, 2009.
Wolf, Susan R. *Meaning in Life*. Princeton, NJ: Princeton University Press, 2010.
Wollgart, Siegfried. 'Zum Wandel von Rationalitätsvorstellungen vom 17. bis zum 20. Jahrhundert.' In *Rationalität heute: Vorstellungen, Wandlungen, Herausforderungen*, edited by Gerhard Banse and Andrzej Kiepas, 15–39. Münster: LIT Verlag, 2002.
Wolterstorff, Nicholas. 'Theology and Science: Listening to Each Other.' In *Religion and Science: History, Method, Dialogue*, edited by W. Mark Richardson and Wesley J. Wildman, 95–104. London: Routledge, 1996.
Wong, Paul T. P. *The Human Quest for Meaning: Theories, Research, and Applications*. Personality and Clinical Psychology Series. 2nd edn. New York: Routledge, 2012.
Wood, Alexander. *In Pursuit of Truth: A Comparative Study in Science and Religion*. London: Student Christian Movement, 1927.
Woodcock, Leslie V. 'Phlogiston Theory and Chemical Revolutions.' *Bulletin of the History of Chemistry* 30, 2 (2005): 63–9.
Woodward, James. *Making Things Happen: A Theory of Causal Explanation*. Oxford: Oxford University Press, 2003.
Woody, Andrea. 'Re-Orienting Discussions of Scientific Explanation: A Functional Perspective.' *Studies in History and Philosophy of Science* 51, 1 (2015): 79–87.

Wootton, David. *The Invention of Science: A New History of the Scientific Revolution*. London: Allen Lane, 2015.
World Health Organization. *International Classification of Functioning, Disability and Health*. Geneva: World Health Organization, 2001.
Worrall, John. 'Fresnel, Poisson and the White Spot: The Role of Successful Predictions in the Acceptance of Scientific Theories.' In *The Uses of Experiment: Studies in the Natural Sciences*, edited by David Gooding, Trevor Pinch, and Simon Schaffer, 135-57. Cambridge: Cambridge University Press, 1989.
Woźniak, Robert J. and Giulio Maspero, eds. *Rethinking Trinitarian Theology: Disputed Questions and Contemporary Issues in Trinitarian Theology*. London: T&T Clark, 2012.
Wright, Cory D. 'The Ontic Conception of Scientific Explanation.' *Studies In History and Philosophy of Science* 54 (2015): 20-30.
Wright, N. T. *Paul and the Faithfulness of God*. 2 vols. London: SPCK, 2013.
Yandell, Keith. *Philosophy of Religion: A Contemporary Introduction*. London: Routledge, 1999.
Ye, Feng. 'Naturalism and Abstract Entities.' *International Studies in the Philosophy of Science* 24 (2010): 129-46.
Zachhuber, Johannes. *Theology as Science in Nineteenth-Century Germany: From F.C. Baur to Ernst Troeltsch*. Oxford: Oxford University Press, 2013.
Zagorin, Perez. 'Francis Bacon's Objectivity and the Idols of the Mind.' *British Journal for the History of Science* 34, 4 (2001): 379-93.
Zahl, Simeon. 'On the Affective Salience of Doctrines.' *Modern Theology* 31, 3 (2015): 428-44.
Zemach, Eddy M. *Real Beauty*. University Park, PA: Pennsylvania State University Press, 1997.
Zollman, Kevin J. S. 'The Communication Structure of Epistemic Communities.' *Philosophy of Science* 74, 5 (2007): 574-87.
Zouboulakis, Michel S. *The Varieties of Economic Rationality: From Adam Smith to Comtemporary Behavioural and Evolutionary Economics*. New York: Routledge, 2014.
Zucca, Lorenzo. 'A New Legal Definition of Religion?' *King's Law Journal* 25, 1 (2014): 5-7.

Index

abductive approaches to investigating reality 175–82
akratic actions 23
d'Alembert, Jean Baptiste le Rond 34–5
Anselm of Canterbury 167–8
Aquinas *see* Thomas Aquinas
Arago, François 120
arguments for the existence of God
  as affirmations of the rationality of Christianity 145–6
  Aquinas's Five Ways 145–7, 167
  Aquinas's Second Way 145–8
  arguments from design 173
  arguments from desire 105–6, 141–3, 180–1
  Kalam argument 169
  ontological arguments 167–8
Aristotle 20, 32, 75, 184
Arianism 108
Athanasius of Alexandria 108
Atkins, Peter 9, 57
Augustine of Hippo 109, 151, 152, 195

Bacon, Francis 111, 157
Barbour, Ian 55, 73
Barth, Karl 141 n. 70, 197
Baxter, Richard 194–5
Bayes's theorem 130, 174
benzene, discovery of ring structure 160
Berger, Peter 76
Bernard, Claude 160
'Beyond Rationality' research programme 40
Bhaskar, Roy 14, 67–8
black swans, and deductive reasoning 165–6
Boff, Leonardo 201
Bohm, David 131
Bohr, Niels 133
Bonaparte, Napoleon 47
bootstrapping, rational 20 n. 9
Bouvard, Alexis 163

Brooke, John Hedley 6–7
Brunner, Emil 196

Cartwright, Nancy 205–6
Castoriadis, Cornelius 25
certainty, craving for, as epistemic vice 36–7
Chalcedon, Council of 208
Clayton, Philip 1–2, 104
'clear and distinct ideas', and rationalism 200, 201–2
cognitive interdisciplinarity 226
Cold War rationality 36
collectivizing of human reason 36–7, 76–80
Collingwood, R. G. 35
common sense 28–31
'conflict' model of science and religion 6, 8–9, 13, 55–6
  foundational role in the 'New Atheism' 9–10
consilience 14, 203–4
Conway Morris, Simon 179–80
Cooper, David E. 200
Copenhagen approach to quantum theory 10, 210
Copernicus, Nicholas 75, 133
Coplestone, Frederick 146
Cottingham, John 200
Coulson, Charles A. 63–5
Craig, William Lane 169–70
Crick, Francis 70
criteria of theory evaluation 101–26
  beauty 117–19
  capacity to predict novel observations 119–22
  coherence, internal 106–9
  correspondence with an external reality 106–9
  elegance 117–19
  objectivity 110–12
  simplicity 113–16
critical realism 14, 67–9

Index

cultural invariance, assumption of 30–1
cultural metanarratives 25–6

d'Alembert, Jean-Baptiste le Rond 34–5
Dalton, John 97
Dante Alighieri 151–2
dark energy 188
Darwin, Charles 75, 97, 98, 101, 121–2, 171–2, 187–8
Dawkins, Richard 138, 185
de Broglie, Louis 131
de Vlamingh, Willem 165
deductive approaches to the investigation of reality 125, 136, 164–9
deductive-nomological approach to scientific explanation 125, 136, 165–7
deism 143, 197
Dennett, Daniel 11
Descartes, René 19, 132, 200
Destutt de Tracy, Antoine Louis Claude 22, 46–7
Dewey, John 4, 13, 46, 152
domain-specific approaches to scientific rationality 38–40
Draper, Paul 218
Duhem, Pierre 40, 144, 163

Eagleton, Terry 128–9
Einstein, Albert 98, 133, 190, 215–16
Einstein-Podolsky-Rosen correlations 131
endoxos, as cultural criterion of rationality 32
Engels, Friedrich 213
English Enlightenment 27–8
Enlightenment, The
    concepts of rationality 3–4, 22, 27, 32, 34–5, 44, 46, 54, 69, 73, 79, 80, 84–5, 88, 203
    definitional problems concerning 33–4
    in England 27–8
    and intellectual colonialism 87–8
epistemic communities 76–9
epistemic dependency and 'big science' 90
epistemic virtues 95–6, 106–10

epistemological decolonialization 88
epistemological pluralism in natural sciences 2–3, 10, 59, 67–8, 143–4, 206, 221
Erklären–Verstehen distinction 30, 126–7, 150–1
Evans-Pritchard, Edward Evan 29
explanation 125–53
    causal approaches 129–32
    deductive-nomological approach 125, 136, 165–7
    difficulties in defining what it means to 'explain' 125–8
    epistemic approaches 135–7, 140–3
    Erklären–Verstehen distinction 30, 126–7, 150–1
    inference to best explanation 101–6
    ontic approaches 135–7, 140–5, 169
    religious approaches 137–53
    self-evidencing explanations 102, 139–40, 144
    unificationist approaches 132–5

falsification and theory testing 114, 159, 162–3, 167
Farrer, Austin 193
Feyerabend, Paul 3 n. 8, 9–10, 125 n. 5
Feynman, Richard 185
Fisher, R. A. 114
Frazer, James George 12, 129
Fresnel, August 120

Gadamer, Hans Georg 79
Galileo, Galilei 65, 133
Gaukroger, Stephen 110
Geertz, Clifford 11, 29
Giere, Ronald J. 61
Gore, Charles 107–8
Grayling, A. C. 158
Gregory, Brad 30

Haidt, Jonathan 197–9
Halley, Edmund 103
Hamman, Johann Georg 77
Hampshire, Stuart 214, 216–17
Hanson, N. R. 176–7
Harman, Gilbert 103
Harrison, Peter 7–8

Index

Heidegger, Martin 29, 79, 112
Heisenberg, Werner 10, 189, 210–11
Heller, Erich 119
Hempel, Carl G. 125, 136, 144, 165–7
Herschel, John 187
Hesse, Mary 159
heuristics and biases research
    programme 25
Hitchens, Christopher 37, 128
Hobbes, Thomas 32–3
Hume, David 85, 131, 169, 171–2
    critique of induction 165, 170–2
Hutton, Frederick W. 121
Huygens, Christiaan 103
Hypothetico-Deductive approach 167

ideology 22–3, 46–9
    as rationalist alternative to
        theology 46
    Marx's critique of 47–8
image of God 148–51
imagination, role of in scientific theory
    development 162, 177, 185
inductive approaches to investigating
    reality 100, 170–5
inference to best explanation, and theory
    choice 101–6
integral theory 14
intelligence, distinguished from
    rationality 42–3
interdisciplinarity 2 n. 3
irrationality 20, 23, 26–8, 35–6, 45–6,
    50, 52, 182–7

Joyce, James 54
Julian of Norwich 145

Kagan, Jerome 41–2
Kalam argument for the existence of
    God 169
Kant, Immanuel 33–4
Kekulé, August 160
Keltner, Decher 197–9
Kepler, Johann 115, 148
Kidd, Ian 57
Kierkegaard, Søren 53
Kitcher, Philip 131, 134
Knorr Cetina, Karin 89
Kornblith, Hilary 96

Kuhn, Thomas 99
Kurtz, Paul 79

Lambda-CDM cosmological
    model 158–9
Latour, Bruno 89, 204
Leibniz, Gottfried Wilhelm 168
Lewis, C. S. 60, 95, 104, 108, 139, 141–3,
    180–1
lex orandi, lex credendi 91–2, 112
Lindbeck, George 108 n. 50
Linnaeus, Carl 134
Lipton, Peter 139–40
Locke, John 19, 77
London School of Economics 'Beyond
    Rationality' programme 40
logic of discovery 141, 159–64, 176
logic of justification 159–64
logical positivism 1
Luther, Martin 52–3, 75, 152
Lycan, William 140

MacIntyre, Alasdair 81–3, 219
Magic 31, 52
Manhattan Project 90
Marcel, Gabriel 192–4
Marx, Karl 47–8, 213
mathematics as torch 132
mathesis universalis 69
Maximus the Confessor 191
Maxwell, James Clerk 132
McGinn, Colin 201
Mendelssohn, Moses 80
'metaphysical nomological pluralism'
    (Cartwright) 205–6
Midgley, Mary 5, 58, 212–13
Mill, John Stuart 120
Millar, Hugh 119
Mitchell, Basil 105
Morin, Edgar 14
multiple levels of real world 65–9
multiple modernities 43–4
multiple perspectives on real
    world 59–65
multiple rationalities 39–41
musement, as theory-generating
    strategy 178–9
mystery
    not equivalent to irrationality 182–6

Index

mystery (cont.)
  in natural sciences 187–90
  as theological category 190–7

Nagel, Thomas 33, 80, 146
natural philosophy 54, 154–5
natural sciences
  criteria of theory choice 106–23
  development of theories 97–106
  as empirical undertakings 4–5, 11–12, 155–7, 159–60, 207
  epistemological pluralism of the sciences 2–3, 10, 59, 67–8, 143–4, 206, 221
  communal aspects of scientific knowledge 89–91
  concepts of scientific explanation 125–37
  and mystery 187–90
natural theology 154
Neurath, Otto 205
'New Atheism' 9–10, 37
Newton, Isaac 24 n. 32, 54, 85, 115, 117, 132, 196
Niebuhr, H. Richard 150
Nietzsche, Friedrich 35, 60–1, 80
numinous, category of 182–3

objectivity, as epistemological virtue 31
Ockham, William of *see* William of Ockham
ontological unity of nature 2, 10, 59, 67–8, 143–4, 206, 221
Oppenheim, Paul 125
Ortega y Gasset, José 22 n. 19
Otto, Rudolf 182–3
Oxford Enzyme Group 90–1

Paley, William 173
Pannenberg, Wolfhart 107
parsimony, principle of 114–15
Peirce, Charles S. 46, 164, 175–8
pensée complexe 14
perspectival realism 61–3
perspectivalism
  in Nietzsche 60–1, 80
  in Plato 60
physico-theology 172–3
Piaget, Jean 198–9

Planck, Max 133
Plato 60
pluralism, epistemological, in natural sciences 2–3, 10, 59, 67–8, 143–4, 206, 221
Poisson, Siméon Denis 120
Pope, Alexander 37
Popper, Karl 114, 122, 138, 152, 160, 162, 207
postmodern-modern binary, inadequacy of 44–5
practice, turn to, in contemporary theory 38
prediction in relation to theory choice 119–22
Prosper of Aquitaine 91
psychology of cognitive interdisciplinarity 226
Putnam, Hilary 107

Quine, W. V. O. 139, 163

Rahner, Karl 197
Ramsey, Ian T. 140
rational virtues and theory choice
  beauty 117–19
  capacity to predict novel observations 119–22
  coherence, internal 106–9
  correspondence with an external reality 106–9
  elegance 117–19
  objectivity 110–12
  simplicity 113–16
rationality
  and akratic actions 23
  changing conceptions of 19–22
  and Cold War 36
  and 'common sense' 28–31
  correlation of rationalities 53–6
  as culturally embedded 23–4, 25–6, 27–31, 41
  Enlightenment notion of 3–4, 22, 27, 32, 34–5, 44, 46, 54, 69, 73, 79, 80, 84–5, 88, 203
  as epistemic 42
  as historically situated 23–4
  and human cognitive processes 25
  and human *sensorium* 24

Index

distinguished from intelligence 42–3
  as instrumental 42
  multiple situated rationalities 2, 14–16, 39–42, 206
  as physically embodied 23–4, 27–31
  'rational' and 'irrational' as social binary 27
  role models of embodied rationality 32–4
  and social imaginaries 25, 46–7, 87
  as socially constructed concepts and practices 41, 76–91
  and social location 75–6
  in social sciences 29–30
  transversal approaches to 45, 73
reasoning, human
  abductive approaches 175–82
  deductive approaches 164–9
  inductive approaches 170–5
  limited capacity to grasp reality 185–6, 200–1, 202
  sufficient reason, principle of 168
religion
  difficulties in defining 10–11, 52–3
  evolutionary theories of origins 72–3
  and magic 52
  science and religion, as interdisciplinary field 50–1, 92
Rescher, Nicholas 28
Robinson, Marilynn 12
Rose, Steven 2, 59–60, 66, 211
Rueger, Alexander 61–2
Ruskin, John 111–12
Russell, Bertrand 35–6, 146

Salmon, Wesley 130, 135
Sartre, Jean-Paul 19 n. 1
Sayers, Dorothy L. 149
Scholz, Heinrich 69
Schrag, Calvin 45 n. 117
scientific method 4–5, 9–10, 11–12, 155–7, 159–60, 207
scientific socialism 212–14
scientism 56–9, 71–2
  as intellectual colonization of the humanities 203–4
  intellectual motivations for 57
searchlight approach to theory development 138

sensorium, human 24
Sergeant, John 77
Simon, Herbert A. 82
Sober, Elliott 116
social imaginary 25, 46–7, 87
Sosa, Ernst 141
Spinoza, Baruch 80
Stein, Edith 152
stratification of reality 65–9
supernatural, as problematic category 218–19
Swinburne, Richard 115, 117, 173–4

Tawney, R. H. 215
Taylor, Charles 25, 28, 84–6
Tertullian, alleged irrationality of 53
Thagard, Paul 103, 172
theology, Christian
  as communal activity 76, 91–2
  distinguished from philosophical deism or theism 14, 98–9, 208–9, 225
  forms of reasoning within 167–70, 172–5, 178–82
  relationship with worship and prayer 91–2, 112
theory, nature of 97–101
theory-laden character of observation 157–9
Thomas Aquinas 75, 167
Tolkien, J. R. R. 108
Torrance, Thomas F. 68
Toulmin, Stephen 80–1
transdisciplinarity 2 n. 3, 204
transversal reasoning 45, 73
Trinity, doctrine of 143–5, 180–1
  as a *theoria* 145
'Two Books' tradition 186–7

under-determination of theory by evidence 102
Uranus, and falsification 163–4

Valcárcel, Vicente Fernández 54
van Fraassen, Bas 129, 190, 207
van Huyssteen, Wentzel 44–5
Vico, Giambattista 186–7
Vienna Circle 207
'view from nowhere' 31, 48, 111, 219

von Baier, Karl Ernst 114
von Helmholtz, Hermann 85

Wainwright, William J. 105
'warfare' model of science and religion 6, 8–9, 13, 55–6
    foundational role in the 'New Atheism' 9–10
Weber, Max 43, 186
Weil, Simone 138
Weinberg, Steven 151
Welsch, Wolfgang 45
Whewell, William 120, 126, 148–9, 211
Whitehead, Alfred North 184–5

Wigner, Eugene 188–9, 204
Wilber, Ken 14, 21 n. 12
William of Ockham 114
Wilson, Edward O. 14, 203–4, 209–10, 225–6
Winch, Peter 29–30
'wise', the 32–5
Wittgenstein, Ludwig 5–7, 20, 21, 79, 137–8
wonder, as a gateway experience 154–5, 184, 190
World Health Organization 69–70
Wright, Sewell G. 114

Yandell, Keith 138